Cyclopedia

of

Engineering

A General Reference Work on

STEAM BOILERS AND PUMPS, STEAM, STATIONARY, LOCOMOTIVE, AND MARINE ENGINES, STEAM TURBINES, GAS AND OIL ENGINES, PRODUCERS, ELEVATORS, HEATING AND VENTILATION, COMPRESSED AIR, REFRIGERATION, TYPES AND MANAGEMENT OF DYNAMO-ELECTRIC MACHINERY, POWER STATIONS, ETC.

Editor-in-Chief

LOUIS DERR, S. B., A. M.

PROFESSOR OF PHYSICS, MASSACHUSETTS INSTITUTE OF TECHNOLOGY

Assisted by

CONSULTING ENGINEERS, TECHNICAL EXPERTS, AND DESIGNERS OF THE HIGHEST PROFESSIONAL STANDING

Illustrated with over Two Thousand Engravings

SEVEN VOLUMES

CHICAGO

AMERICAN TECHNICAL SOCIETY

1915

Editor-in-Chief

LOUIS DERR, S. B., A. M.

Professor of Physics, Massachusetts Institute of Technology

Authors and Collaborators

LIONEL S. MARKS, S. B., M. M. E.

Professor of Mechanical Engineering, in Harvard University and Massachusetts Institute of Technology
American Society of Mechanical Engineers

LLEWELLYN V. LUDY, M. E.

Professor of Experimental Engineering, Purdue University
American Society of Mechanical Engineers

LUCIUS I. WIGHTMAN, E. E.

Consulting Engineer and Counsellor in Technical Advertising, New York City

FRANCIS B. CROCKER, E. M., Ph. D.

Professor of Electrical Engineering, Columbia University, New York
Past President, American Institute of Electrical Engineers

W. H. FREEDMAN, C. E., E. E.

Head, Department of Electrical Engineering, University of Vermont

GEORGE C. SHAAD, E. E.

Professor of Electrical Engineering, University of Kansas

WALTER S. LELAND, S. B.

Formerly Assistant Professor of Naval Architecture, Massachusetts Institute of Technology
American Society of Naval Architects and Marine Engineers

Authors and Collaborators—Continued

CHARLES L. HUBBARD, S. B., M. E.

Consulting Engineer on Heating, Ventilating, Lighting, and Power

ARTHUR L. RICE, M. M. E.

Editor, *The Practical Engineer*

ROBERT H. KUSS, M. E.

Sales Manager, Edge Moor Iron Company, Chicago
American Society Mechanical Engineers

HENRY H. NORRIS, M. E.

Formerly Professor and Acting Head, Department of Electrical Engineering, Cornell University

H. S. McDEWELL, S. B., M. M. E.

Instructor in Mechanical Engineering, University of Illinois
Formerly Gas Engine Erection Engineer, Allis-Chalmers Manufacturing Company, Milwaukee, Wisconsin
American Society of Mechanical Engineers

GLENN M. HOBBS, Ph. D.

Secretary and Educational Director, American School of Correspondence
Formerly Instructor in Physics, University of Chicago
American Physical Society

LOUIS DERR, S. B., A. M.

Professor of Physics, Massachusetts Institute of Technology

JOHN H. JALLINGS

Mechanical Engineer and Elevator Expert
For Twenty Years Superintendent and Chief Constructor for J. W. Reedy Elevator Company

Authors and Collaborators—Continued

A. H. NUCKOLLS, B. A.
Chemical Engineer, Underwriters' Laboratories, Inc.

MILTON W. ARROWOOD
Graduate, United States Naval Academy
Refrigerating and Mechanical Engineer
Consulting Engineer

C. C. ADAMS, B. S.
Switchboard Engineer with General Electric Company

CHESTER A. GAUSS, E. E.
Formerly Associate Editor, "Electrical Review and Western Electrician"

EDWARD B. WAITE
Dean and Head, Consulting Department, American School of Correspondence
Formerly Research Assistant, Harvard College Observatory
American Society of Mechanical Engineers; Western Society of Engineers

J. P. SCHROETER
Consulting Engineer, Electrical Engineering Department, American School of Correspondence

WILLIAM S. NEWELL, S. B.
With Bath Iron Works
Formerly Instructor, Massachusetts Institute of Technology

CARL S. DOW, S. B.
With Walter B. Snow, Publicity Engineer, Boston
American Society of Mechanical Engineers

JESSIE M. SHEPHERD, A. B.
Head, Publication Department, American School of Correspondence

Authorities Consulted

THE editors have freely consulted the standard technical literature of Europe and America in the preparation of these volumes. They desire to express their indebtedness particularly to the following eminent authorities, whose well-known treatises should be in the library of every engineer.

Grateful acknowledgment is made here also for the invaluable co-operation of the foremost engineering firms in making these volumes thoroughly representative of the best and latest practice in the design and construction of steam and electrical machines; also for the valuable drawings and data, suggestions, criticisms, and other courtesies.

JAMES AMBROSE MOYER, S. B., A. M.

Member of The American Society of Mechanical Engineers; American Institute of Electrical Engineers, etc.; Engineer, Westinghouse, Church, Kerr and Company
Author of "The Steam Turbine," etc.

E. G. CONSTANTINE

Member of the Institution of Mechanical Engineers; Associate Member of the Institution of Civil Engineers
Author of "Marine Engineers"

C. W. MacCORD, A. M.

Professor of Mechanical Drawing, Stevens Institute of Technology
Author of "Movement of Slide Valves by Eccentrics"

CECIL H. PEABODY, S. B.

Professor of Marine Engineering and Naval Architecture, Massachusetts Institute of Technology
Author of "Thermodynamics of the Steam Engine," "Tables of the Properties of Saturated Steam," "Valve Gears to Steam Engines," etc.

FRANCIS BACON CROCKER, M. E., Ph. D.

Head of Department of Electrical Engineering, Columbia University; Past President American Institute of Electrical Engineers
Author of "Electric Lighting," "Practical Management of Dynamos and Motors"

SAMUEL S. WYER

Mechanical Engineer; American Society of Mechanical Engineers
Author of "Treatise on Producer Gas and Gas-Producers," "Catechism on Producer Gas"

E. W. ROBERTS, M. E.

Member, American Society of Mechanical Engineers
Author of "Gas-Engine Handbook," "Gas Engines and Their Troubles," "The Automobile Pocket-Book," etc.

Authorities Consulted—Continued

GARDNER D. HISCOX, M. E.

Author of "Compressed Air," "Gas, Gasoline, and Oil Engines," "Mechanical Movements," "Horseless Vehicles, Automobiles, and Motorcycles," "Hydraulic Engineering," "Modern Steam Engineering," etc.

EDWARD F. MILLER

Professor of Steam Engineering, Massachusetts Institute of Technology
Author of "Steam Boilers"

ROBERT M. NEILSON

Associate Member, Institution of Mechanical Engineers; Member of Cleveland Institution of Engineers; Chief of the Technical Department of Richardsons, Westgarth and Company, Ltd.
Author of "The Steam Turbine"

ROBERT WILSON

Author of "Treatise on Steam Boilers," "Boiler and Factory Chimneys," etc.

CHARLES PROTEUS STEINMETZ

Consulting Engineer, with the General Electric Company; Professor of Electrical Engineering, Union College
Author of "The Theory and Calculation of Alternating-Current Phenomena," "Theoretical Elements of Electrical Engineering," etc.

JAMES J. LAWLER

Author of "Modern Plumbing, Steam and Hot-Water Heating"

WILLIAM F. DURAND, Ph. D.

Professor of Marine Engineering, Cornell University
Author of "Resistance and Propulsion of Ships," "Practical Marine Engineering"

HORATIO A. FOSTER

Member, American Institute of Electrical Engineers; American Society of Mechanical Engineers; Consulting Engineer
Author of "Electrical Engineer's Pocket-Book"

ROBERT GRIMSHAW, M. E.

Author of "Steam Engine Catechism," "Boiler Catechism," "Locomotive Catechism," "Engine Runners' Catechism," "Shop Kinks," etc.

SCHUYLER S. WHEELER, D. Sc.

Electrical Expert of the Board of Electrical Control, New York City; Member American Societies of Civil and Mechanical Engineers
Author of "Practical Management of Dynamos and Motors"

Authorities Consulted—Continued

J. A. EWING, C. B., LL. D., F. R. S.

Member, Institute of Civil Engineers; formerly Professor of Mechanism and Applied Mechanics in the University of Cambridge; Director of Naval Education
Author of "The Mechanical Production of Cold," "The Steam Engine and Other Heat Engines"

LESTER G. FRENCH, S. B.

Mechanical Engineer
Author of "Steam Turbines"

ROLLA C. CARPENTER, M. S., C. E., M. M. E.

Professor of Experimental Engineering, Cornell University; Member, American Society of Heating and Ventilating Engineers; Member, American Society of Mechanical Engineers
Author of "Heating and Ventilating Buildings"

J. E. SIEBEL

Director, Zymotechnic Institute, Chicago
Author of "Compend of Mechanical Refrigeration"

WILLIAM KENT, M. E.

Consulting Engineer; Member, American Society of Mechanical Engineers, etc.
Author of "Strength of Materials," "Mechanical Engineer's Pocket-Book," etc.

WILLIAM M. BARR

Member, American Society of Mechanical Engineers
Author of "Boilers and Furnaces," "Pumping Machinery," "Chimneys of Brick and Metal," etc.

WILLIAM RIPPER

Professor of Mechanical Engineering in the Sheffield Technical School; Member, The Institute of Mechanical Engineers
Author of "Machine Drawing and Design," "Practical Chemistry," "Steam," etc.

J. FISHER-HINNEN

Late Chief of the Drawing Department at the Oerlikon Works
Author of "Continuous Current Dynamos"

SYLVANUS P. THOMPSON, D. Sc., B. A., F. R. S., F. R. A. S.

Principal and Professor of Physics in the City and Guilds of London Technical College
Author of "Electricity and Magnetism," "Dynamo-Electric Machinery," etc.

ROBERT H. THURSTON, C. E., Ph. B., A. M., LL. D.

Director of Sibley College, Cornell University
Author of "Manual of the Steam Engine," "Manual of Steam Boilers," "History of the Steam Engine," etc.

Authorities Consulted—Continued

JOSEPH G. BRANCH, B. S., M. E.

Chief of the Department of Inspection, Boilers and Elevators; Member of the Board of Examining Engineers for the City of St. Louis

Author of "Stationary Engineering," "Heat and Light from Municipal and Other Waste," etc.

JOSHUA ROSE, M. E.

Author of "Mechanical Drawing Self Taught," "Modern Steam Engineering," "Steam Boilers," "The Slide Valve," "Pattern Maker's Assistant," "Complete Machinist," etc.

CHARLES H. INNES, M. A.

Lecturer on Engineering at Rutherford College

Author of "Air Compressors and Blowing Engines," "Problems in Machine Design," "Centrifugal Pumps, Turbines, and Water Motors," etc.

GEORGE C. V. HOLMES

Whitworth Scholar; Secretary of the Institute of Naval Architects, etc.

Author of "The Steam Engine"

FREDERIC REMSEN HUTTON, E. M., Ph. D.

Emeritus Professor of Mechanical Engineering in Columbia University; Past Secretary and President of American Society of Mechanical Engineers

Author of "The Gas Engine," "Mechanical Engineering of Power Plants," etc.

MAURICE A. OUDIN, M. S.

Member of American Institute of Electrical Engineers

Author of "Standard Polyphase Apparatus and Systems"

WILLIAM JOHN MACQUORN RANKINE, LL. D., F. R. S. S.

Civil Engineer; Late Regius Professor of Civil Engineering in University of Glasgow

Author of "Applied Mechanics," "The Steam Engine," "Civil Engineering," "Useful Rules and Tables," "Machinery and Mill Work," "A Mechanical Textbook"

DUGALD C. JACKSON, C. E.

Head of Department of Electrical Engineering, Massachusetts Institute of Technology

Member of American Institute of Electrical Engineers

Author of "A Textbook on Electro-Magnetism and the Construction of Dynamos," "Alternating Currents and Alternating-Current Machinery"

A. E. SEATON

Author of "A Manual of Marine Engineering"

WILLIAM C. UNWIN, F. R. S., M. Inst. C. E.

Professor of Civil and Mechanical Engineering, Central Technical College, City and Guilds of London Institute, etc.

Author of "Machine Design," "The Development and Transmission of Power," etc.

ENGINE-DRIVEN MULTIPHASE ALTERNATOR WITH DIRECT CONNECTED EXCITER
Courtesy of Fort Wayne Electric Works

Foreword

THE "prime mover", whether it be a massive, majestic Corliss, a rapidly rotating steam turbine, or an iron "greyhound" drawing the Limited, is a work of mechanical art which commands the admiration of everyone. And yet, the complicated mechanisms are so efficiently designed and everything works so noiselessly, that we lose sight of the wonderful theoretical and mechanical development which was necessary to bring these machines to their present state of perfection. Notwithstanding the genius of Watt, which was so great that his basic conception of the steam engine and many of his inventions in connection with it exist today practically as he gave them to the world over a hundred years ago, yet the mechanics of his time could not build engine cylinders nearer true than three-eighths of an inch — the error in the modern engine cylinders must not be greater than two-thousandths of an inch.

¶ But the developments did not stop with Watt. The little refinements brought about by the careful study of the theory of the heat engine; the reduction in heat losses; the use of superheated steam; the idea of compound expansion; the development of the Stephenson and Walschaert valve gears — all have contributed toward making the steam engine almost mechanically perfect and as efficient as is inherently possible.

¶ The development of the steam turbine within recent years has opened up a new field of engineering, and the adoption of this form of prime mover in so many stationary plants like the immense Fisk Station of the Commonwealth Edison Company, as well as its use on the gigantic ocean liners like the Lusitania, makes this angle of steam engineering of especial interest.

¶ Adding to this the wonderful advance in the gas engine field — not only in the automobile type where requirements of lightness, speed, and reliability under trying conditions have developed a most perfect mechanism, but in the stationary type which has so many fields of application in competition with its steam-driven brother as well as in fields where the latter can not be of service — you have a brief survey of the almost unprecedented development in this most fascinating branch of Engineering.

¶ This story has been developed in these volumes from the historical standpoint and along sound theoretical and practical lines. It is absorbingly interesting and instructive to the stationary engineer and also to all who wish to follow modern engineering development. The formulas of higher mathematics have been avoided as far as possible, and every care has been exercised to elucidate the text by abundant and appropriate illustrations.

¶ The Cyclopedia has been compiled with the idea of making it a work thoroughly technical, yet easily comprehendible by the man who has but little time in which to acquaint himself with the fundamental branches of practical engineering. If, therefore, it should benefit any of the large number of workers who need, yet lack, technical training, the publishers will feel that its mission has been accomplished.

¶ Grateful acknowledgment is due the corps of authors and collaborators — engineers and designers of wide practical experience, and teachers of well-recognized ability — without whose co-operation this work would have been impossible.

Table of Contents

VOLUME VII

---

* For page numbers, see foot of pages.

† For professional standing of authors, see list of Authors and Collaborators at front of volume.

MANAGEMENT OF DYNAMO-ELECTRIC MACHINERY

PART I

The object of this treatise is to set forth the most important features which must be considered in the actual handling and operation of electric generators and motors. The principles and general construction of direct-current (d.c.) and alternating-current (a.c.) generators and motors are treated elsewhere. The subject may be divided into three parts, as follows:

(1) Selection, Erection, Connection, and Operation
(2) Inspection and Testing
(3) Troubles, or "Diseases," and Remedies.

THE MACHINE—FROM SELECTION TO SERVICE

SELECTION OF A MACHINE

The voltage, capacity, and type of machine are dependent upon the system to which it is to be connected and the purpose for which it is to be utilized, but there are certain general features which should be considered in every case.

Construction. The construction should be of the most solid character and guaranteed first-class in every respect, including materials and workmanship.

Finish. A good finish is desirable, since it is likely to cause the attendant to take greater care of the equipment.

Simplicity. The machine should be as simple as possible in all its parts; peculiar or complicated features should be avoided, unless absolutely essential for the operation of the system.

Attention. The amount of attention required by a machine depends in a great measure on the kind of machine. Direct-connected revolving-field a.c. generators, revolving-field synchronous motors, and induction motors require little or no attention, whereas machines of the commutating type demand a certain amount of care to keep brushes and commutators in condition. Above all, cleanliness is essential to the life and proper operation of all electrical apparatus.

Handling. Machines of the revolving-field alternating-current type are arranged so that the armatures may be moved on the foundation plates parallel to the shaft to allow of repairs to either field or armatures. Direct-current types of machines are usually split through the magnet frame in a horizontal plane so that they may be disassembled readily.

Regulation. Unless the machine is self-regulating, it is usually customary to employ some automatic device external to the machine to maintain constant voltage.

Form. Preferably the machine should be of standard reliable make, to insure proper symmetry of form as well as ruggedness of design.

Capacity. A machine should be carefully selected for the work it has to perform, the highest efficiency being obtained at normal and slight overloads. The heating at normal and prescribed overloads is fixed by the Standardization Rules of the American Institute of Electrical Engineers.

Cost. It is usually an error to select a generator or motor simply because it is cheap, since both the materials and the workmanship required for the construction of a high-grade electrical machine are costly.

ERECTION

MECHANICAL CONDITIONS

Location. The place chosen for the machine should, if possible, be *dry, free from dust or grit, light, and well ventilated.* It should also be arranged so that there is room enough for the removal of the armature without shifting or turning the machine.

Foundations. It is of great importance to have the machine firmly placed upon a good and solid foundation; otherwise, no matter how well constructed and well managed, the vibrations occurring on a poor foundation will produce noise, sparking at the brushes, and other troubles.

It is also necessary, if the machine is belt-driven, to mount it upon rails or a sliding bed-plate provided with holding-down bolts and tightening screws for aligning and adjusting the belt while the machine is in operation, Fig. 1. The machinery foundations consist of a mass of stone, masonry, brickwork, or concrete, upon which the machinery is placed and usually held firmly in place by bolts passing entirely through the mass. These bolts are built into the foundations, the proper positions for them being determined by a wood template suspended above the foundation,

Fig. 1. General Electric Belted Generator Showing Adjustable Base

as shown in Fig. 2. The bolts are preferably surrounded by iron pipe that fixes them longitudinally but allows a little side play which may be necessary to enable them to enter the bed-plate holes readily. The brickwork for machinery foundations should consist of hard burned bricks of first quality, *laid in good cement mortar. Lime mortar is entirely unfit for the purpose,* being likely to crumble away under the effect of the vibrations caused by the machinery. Brick or concrete foundations should be finished with a cap of cement. This forms a level surface upon which to set the machinery. Almost all large foundations are concrete, brick seldom being used.

Fixing the Machine. *Levelling.* In fixing either direct-connected or belt-driven machines, determine, with a long straight edge and spirit level, checked by a surveyor's Y level, if the top of the foundation is level and true. If this is found to be the case, the holding-down bolts may be dropped into the holes in the foundation, if they are not already built in, and the machine carefully placed thereon, the ends of the bolts being passed through the holes in the bed-plate and secured by a few turns of the nuts.

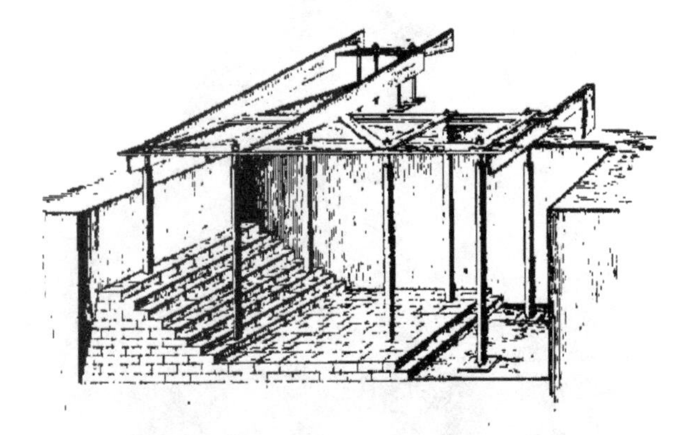

Fig. 2. Diagram of Power House Foundation Showing Wood Template Suspended above Foundation for Construction Purposes

Aligning. The machine should then, if belt-connected, be carefully aligned with the transmitting pulley or flywheel. Particular attention should be paid to the alignment of the pulleys in order that the belt may run properly. If direct-connected, the dynamo bed-plate and armature shaft must be carefully aligned and adjusted with respect to the engine shaft, raising or lowering the bed-plates of the corresponding machines by means of thin cast iron or other wedges; and the generator frame also should be adjusted to its proper height by means of thin strips of metal set between its supporting feet and the bed-plate.

Grouting the Base. Having thus aligned and leveled the machine, it should next be grouted with thin cement. This is done by arranging a wall of mud or wood battens around the bed-plates of the machine, and running in thin cement until the holding-down bolt holes are filled, and the cement has risen at least one-half inch above the under side of the bed-plate. This head of grouting allows for shrinkage as the water dries and the grouting settles. If grouting were poured only to the under side of the bed-plate, it would shrink away when dry and leave a space between the foundation and the bed-plate. When the cement has partially set, the wall may be taken down and the surplus cement removed. When the cement is dry, the nuts on the holding-down bolts may be drawn up. This fixes the machine firmly upon its foundation.

Mechanical Connections

Various means are employed to connect the engine or other prime mover with the generator, or the motor with the apparatus to be driven. The most important are as follows:

Direct Connection	Rope Driving
Belting	Toothed Gearing

Other apparatus, such as shafting, clutches, hangers, and pulleys, are used in combination with the above means.

Direct Connection. Direct connection is the simplest and, for that and other reasons, the most desirable means of connection, provided it can be carried out without involving sacrifices that offset its advantages. This method, also called direct coupling or direct driving, compels the engine and the generator to run at the same speed, which gives rise to some difficulty, as the most desirable speeds of the two machines do not usually agree. The natural speed of a generator is high, while that of a reciprocating engine is low; hence to obtain the same voltage from a direct-connected generator, more conductors are necessary, or the flux cut must be increased. Accordingly, the armature and the frame of the direct-connected generator must be larger, thus making it more expensive than a belt-driven machine. On the other hand the speed of steam turbines is so high that it usually requires specially designed and constructed generators for direct connection.

Continuous Shaft for Engine and Generator. The direct connection of an engine and generator is accomplished in several ways, the simplest of which consists in mounting the armature of the generator directly on one end of the shaft of the engine. This may be accomplished in any one of several ways. Fig. 3 represents a three-bearing belted generator.

Coupled Shafts. Another form of direct coupling is that in which a motor and a generator, each complete in itself, and each having two bearings, are coupled together by some mechanical device, which may be either rigid or slightly elastic or adjustable.

Fig. 3. Fort Wayne Three-Bearing Belted Generator.

In the former case the two shafts are practically equivalent to a single one, which, while making it easy to remove either machine for repairs, is somewhat objectionable owing to the fact that it requires larger foundations and introduces the difficulty of accurately aligning four bearings. The use of a flexible coupling avoids the necessity of perfect alignment, and also the serious trouble that might arise if the settling or the wear of the bearings should be uneven. There are various forms of flexible coupling, one form having rubber cylinders interposed between the two parts.

The direct coupling of generators with hydraulic turbines can usually be carried out without departing from the natural speed of either machine, since the ordinary speed of a turbine agrees closely with the normal speed of a generator of the corresponding capacity.

The relative efficiency of direct coupling and belting depends greatly upon the conditions; but in general the former is more efficient at or near rated load, and the latter at light loads. *The simplicity, compactness, and positive as well as noiseless action* of direct connection have caused it to become the most approved method.

Belting. *Kind.* If the generator or motor is not directly connected—the former to the prime mover, the latter to the apparatus to be driven—it is usually connected by some form of belting. The kind of belting selected depends greatly upon conditions of drive, distances, etc., and it may be leather, rawhide, rubber, or rope. For the ordinary short drives, leather is the most desirable, though, when the power to be transmitted is small, rawhide belts are also satisfactory, especially as the cost is less than for leather belts. For considerable distances, rope driving answers very well, because it is so much lighter and cheaper than an equivalent leather belt, though grooved pulleys are required, making the total cost about the same. Rubber belts are used to advantage in driving generators from water turbines, where the belt might be exposed to moisture. Leather belting is usually the most reliable and satisfactory for general application, except for very short drives, where a form of chain belt works best. There are three thicknesses of leather belting: single, light-double, and double. For use in connection with generators, motors, or other high-speed machinery, the "light-double" belting is usually the best.

Power Transmitted by Belt. The exact amount of power that a given belt is capable of transmitting is not very definite. The ordinary rule is that "single" belt will transmit one horsepower for each inch of its width when traveling at a speed of 1000 feet per minute. If the speed is greater or less, the power is proportionately increased or decreased. This statement of h.p. transmitted is based upon the condition that the belt is in contact with the transmitting pulley around one-half of its circumference, or 180 degrees, which is usually the case. If the arc of contact is less than 180 degrees, the power transmitted is less in the following proportion: An arc of 185 degrees gives 84 per cent, while 90 degrees contact gives only 64 per cent of the power derived from a belt contact of 180 degrees. If, on the other hand, the upper side sags downward, which is always desirable, the belt is in contact with more than half the circumference of the pulley; and thus the grip is considerably increased and more power can be transmitted. These facts make it very desirable to have the *loose side of the belt on top.* If the loose side is below, it sags away from the pulley and is also likely to strike the floor.

An approximate expression for determining the width of a single belt required to transmit a given horsepower is as follows:

$$W = \frac{\text{h.p.} \times 1000}{S \times C}$$

in which W is the width of the belt in inches; h. p. the horsepower to be transmitted; S is the speed of the belt in feet per minute, which is equal to the circumference of the driving pulley in feet multiplied by the number of revolutions per minute; and C is a factor dependent upon the arc of contact.*

"Double" belting is expected to transmit 1½, and "light-double" 1¼ times as much power as "single" belting of the same width. Belting formulas are only approximate and should not be applied too rigidly, since the grip of the belt upon the pulley varies considerably under different conditions of tension, temperature, and moisture. The smooth side of a belt should always be run against a pulley, as it transmits more power and is more durable.

Fig. 4. Belt Clamp for Splicing

Spliced Joints. Belting used for electric machinery, being usually high-speed, should be made "endless" for permanent work, as this makes less noise; but it may be used with laced joints temporarily. A spliced or "endless" joint is made

* Belts slip or "creep" on the pulley about 2 per cent; hence, in determining the size of pulleys whose speed must be accurate, the calculated belt speed should be about 2 per cent too high.

as follows: The two ends of the belt are pared down on opposite sides with a sharp knife into the form of long thin wedges, so that when laid together a long uniform joint is obtained of the *same thickness as the belt itself.* The parts are then firmly joined with cement and sometimes with rivets also. It may be necessary to splice or lace a belt while in position on the pulleys; and for this purpose some form of belt clamp, Fig. 4, should be employed.

If a belt is ordered endless, or is spliced away from the pulleys, great care should be exercised in determining the exact length required. A string that will not stretch, or preferably a wire put around the pulleys in the position to be occupied by the belt, is the best way to avoid a mistake. In measuring for a belt, the generator or motor should be moved on its sliding base so as to make the distance between shaft centers a minimum, in order

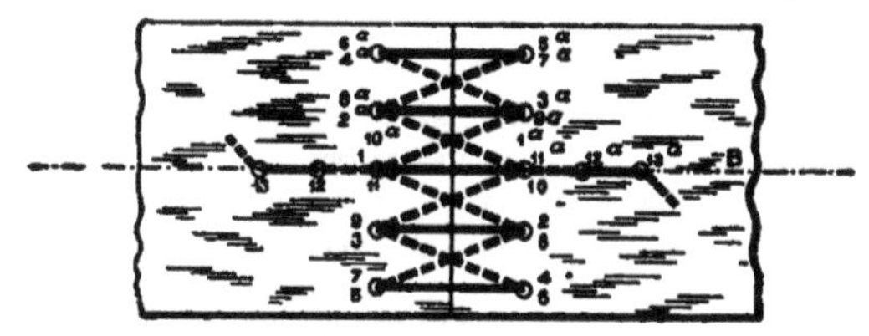

Fig. 5. Diagram Showing Proper Way of Lacing Belt

to allow for the stretch of the belt, which may be as much as one-half inch per foot of length.

Laced Joints. The lacing of a belt is a very simple and common method of making a joint; but it should not be permanently employed at high speeds for electric machinery belting, as it is liable to pound on the pulleys, producing noise, vibration, and sparking; with lighting generators it is also likely to cause flickering in the lamps. In lacing belts, the ends should be cut *perfectly square, and there should be as many stitches of the lace slanting to the left as there are to the right;* otherwise the ends of the belt will shift sidewise, owing to the unequal strain, and the projecting corners may strike or catch in the clothing of persons. A good way to accomplish this is shown in Fig. 5. The various holes should be made with a circular punch, the nearest one being about three-fourths inch from the side, and the line through the center of the row of holes about one inch from the end of the belt. In large belts these distances should be a little greater. A regular belt lacing of strong pliable leather or a special wire should be used. The lacing is doubled to find its middle; and the ends are passed through the two holes marked *1* and *1a*, precisely as in lacing a shoe. The two ends are then passed successively through the two series of holes, in the order in which they are numbered, *2*, *3*, *4*, etc., and *2a*, *3a*, *4a*, etc., finishing at *13* and *13a*, which are additional holes for securing the ends of the lace. The great advantage of this method of lacing is that the lace lies on the pulley side parallel to the direction of motion.

Arrangement and Care of Belting. It is desirable, for satisfactory running, that belts should be reasonably long and nearly horizontal. The distance between centers of two belt-connected pulleys should, if possible, be not less than three times the diameter of the larger pulley. The belt should be just tight enough to avoid slipping, without straining the shaft or bearings. The two shafts which are to be belt-connected must be perfectly parallel, and the centers of the faces of the driving and driven pulleys must be exactly opposite to each other, in a straight line perpendicular to the axis of the shafts. The machines should then be turned over slowly with the belt on, to see if the latter tends to run to one side of the pulley, which would show that it is not yet properly "lined up," in which case one or both machines should be slightly shifted, until the belt runs true. If possible, the machine and the belt should be set and adjusted so as to cause the armature to move back and forth in the bearings while running, on account of the side motion of the belt, and thus make the commutator wear more smoothly, and distribute the oil in the bearings. Where it is impracticable to have sufficient distance between centers of belt-connected pulleys, an idler pulley is often used to give greater arc of contact and elastic tension.

It is always desirable to have belts as pliable as possible; hence the occasional use of a good belt dressing—as neatsfoot oil, etc.,—is recommended. Rosin and other sticky substances are sometimes applied to increase the adhesion; but this is a practice allowable only in an emergency, as it may destroy the belt surface.

In places where the belting is very much exposed, and liable to catch in the clothing, it is advisable to surround it by a railing or box.

Rope Driving. Rope driving possesses advantages over ordinary belting in some cases. The rope runs in V-shaped grooves in the peripheries of the pulleys, and thereby obtains a good grip by a sort of wedging action. The kinds of rope ordinarily employed for this purpose are cotton, hemp, rawhide, and wire.

Advantages. The general advantages are as follows:

(1) Economy in first cost.

(2) Large amount of power that can be transmitted with a given diameter and width of pulley, on account of the grip obtained.

(3) It is almost noiseless.

(4) Ropes, on account of their lightness, can be used to transmit power over greater distances than are possible with any other form of belting; and also for very short distances on account of the wedging action. Manila rope is generally used in the United States, being of three strands, hawser laid, and may be from ½ inch to 2 inches in diameter. The breaking strength varies from 7000 to 12,000 pounds per square inch of cross section. It has been found that the best results are obtained when the tension in the driving side of the rope is only 3 to 4 per cent of the breaking strength.

Power Transmitted by Rope Drive. The diameter of a single rope necessary to transmit a required h.p. is given by the formula:

$$D^2 = \frac{825 \text{ h.p.}}{V\left(200 - \frac{V^2}{1072}\right)}$$

in which h.p. is horsepower transmitted; V is velocity of rope in feet per second; and D is diameter of rope in inches.

The maximum power is obtained at a speed of about 84 feet per second. With higher speeds the centrifugal force becomes so great that the power transmitted decreases rapidly, and at about 142 feet per second it counteracts the whole allowable tension ($200\ D^2$ pounds) and no power is transmitted.

Arrangement of Rope Belting. There are two methods of arranging rope transmission: one consists in using several separate belts; and the other employs a single endless rope which

Table I

Horsepower Transmitted by Cold-Rolled Shafting at Various Speeds

$$\text{h.p.} = \frac{R\,d^3}{100}$$ in which R is revolutions per minute and d is diameter of shaft in inches.

Diameter in Inches	Revolutions per Minute			
	100	200	300	400
1½	3.4	6.7	10	13.5
1⅝	4.3	8.6	12.8	17
1¾	5.4	10.6	16	21
1 15/16	7.3	14.5	22	29
2	8	16	24	31.9
2⅛	9.6	19.1	29	38
2¼	11.4	23	34	45
3	27	54	81	108
3¼	34	68	103	136
3½	43	86	129	171
4	64	128	192	255
5	125	250	375	500

For h.p. transmitted by turned-steel shafting multiply above figures by 0.8.

For ordinary line shafting with supporting bearings about 8 feet apart—multiply by 1.43 for cold-rolled and by 1.11 for turned shafts.

Usual speed of shafting:

Machine Shops........................120 to 250 r.p.m.
Wood Shops........................250 to 300 r.p.m.
Textile Mills........................300 to 400 r.p.m.

passes spirally around the pulley several times and is brought back to the first groove by a slanting idle pulley, and therefore is called the "wound" system. The separate ropes do not require the carrying-over pulley and, if one rope breaks, those remaining are sufficient to transmit the power temporarily; whereas an accident with the single-rope system entirely interrupts the service. In the "multi-rope" system it is practically impossible to make and maintain the belts of exactly equal length, hence the tensions on the various ropes differ, and they hang at different heights on the slack side, producing an awkward appearance.

Toothed Gearing. Toothed gearing possesses the decided advantages of positive action and the ability to give large ratios of speed and small side pressure on the bearings. Nevertheless it is seldom employed for driving generators. The most extensive applications of gearing for electrical purposes are in connection with railway motors, and many industrial applications of a. c. and d. c. motors.

Shafting. An intermediate or countershaft is not desirable since it increases the complication and the frictional losses of the system; but it is often necessary in the generation or application of electric power, either to obtain a greater multiplication of speed than is possible by belting directly, or to enable a single engine or motor to drive a greater number of machines.

The two important kinds of shaftings are "cold-rolled" and "turned". The former is rolled to the exact size and requires no further treatment. It has the advantage of a smooth hard surface, but it is difficult to make it perfectly true and straight. Turned-steel shafting is most commonly employed, and has the advantage that shoulders, journals, or other variations in size can easily be made on it. Table I gives the ordinary data for shafting.

ASSEMBLING OF THE MACHINE

Numbers of Parts and Drawings the Guide. The proper method of setting the foundation plate, or base, of a machine has already been described. The successive parts of the machine should then be assembled in their respective order. The bearing pedestals will usually be found to have a number stamped on them with a corresponding number stamped on the base. The same is true for the magnet frame. Approved drawings showing the correct mechanical assembly and proper electrical connection should be furnished with each machine.

Care of Windings. In handling the parts of the machine, care must be used not to damage the windings in any way. In the case of a direct-current machine, the commutator must be properly protected. If the machine is assembled by means of a crane, the lifting cable shculd not come in contact with any part of the windings. Often a wood spreader, Fig. 6, between turns of the cable is necessary to keep it clear of the windings.

Attention to Joints and Metal Surfaces. All magnetic joints must be made clean and bright. A very light coating of oil will prevent such joints from rusting. The same precaution should apply to all finished surfaces about the machine. The bearings should be carefully examined to see that their bearing surfaces have not been bruised or roughened. Should such defects show up, the bearings should be scraped by an *experienced machinist.* If the bearing surface of the shaft has become bruised or pitted by rust, this should be carefully smoothed before assembling the shaft in the bearings. The oil bearings should be flushed out with gasoline or some light grade of oil before being filled with a good

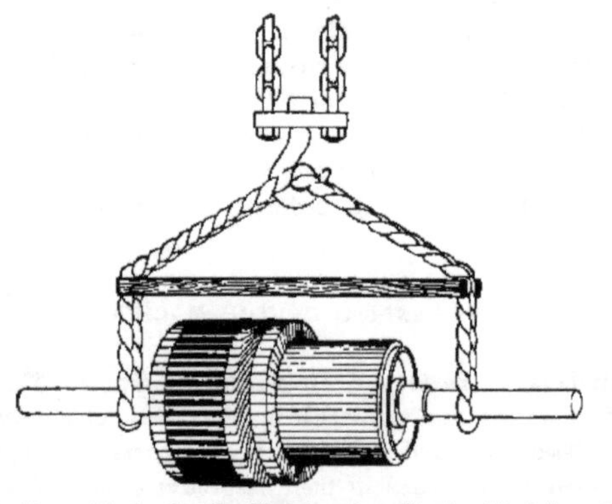

Fig. 6. Proper Rigging for Handling Armature Showing Wood Spreader to Protect Windings

grade of dynamo oil. The field frame should be set so as to obtain a uniform air gap all around.

Brushes and Commutator on D.C. Machines. In the case of a direct-current machine the accurate spacing and setting of the brushes on the commutator is essential to good commutation. Brushes should be fitted to the commutator by drawing a piece of sandpaper under them; then they should be spaced around the commutator by laying off on a thin steel tape or a piece of paper equidistant positions corresponding to the number of poles on the machine. The steel tape is placed around the commutator and either the heel or toe of the brushes on each brush stud made to conform to the marking on the tape. At the same time the brushes of each set must be placed on the commutator parallel to the commutator bars.

Examination Before Starting. Before starting a machine it should be carefully examined to see that there are no loose pieces of metal or other material about the machine. See that the bearings are filled with oil to the proper level. If the machine is belt driven, see that the belt has the proper tension.

Starting a Generator. *Gradual Building-Up of Speed.* In the case of a generator direct-connected to the prime mover, bring up the speed of the set gradually. Be sure that everything is running smoothly and that the oil rings in the self-oiling bearings are turning. The speed may finally be brought up to normal. If it has been determined by insulation test that the machine requires no drying out, the voltage may be built up to normal gradually, and the machine will be ready for regular service.

Drying Out. Large direct-current generators often require drying out before putting in service. This may be accomplished by short circuiting the armature leads beyond the ammeter and "bucking" the shunt and series fields. In the case of an alternator, the phases may be short circuited on the outside of the switchboard instruments so that a reading may be obtained. The field will require very slight excitation to produce full load current with phases short circuited.

Insulation Resistance Test. The insulation resistance should be measured and, if this is found to be the proper amount for the machine under consideration, it may be put into service. If no direct reading instrument is obtainable, the resistance may be measured by using a direct-current voltmeter of known high resistance on a 500 volt direct-current circuit. The method is as follows:

Using a constant potential of about 500 volts, the voltage reading is determined by connecting the terminals of the supply circuit directly to the meter. The insulation resistance to be measured is then connected in series with the voltmeter and a second reading is made and noted. The resistance X is then given by the formula

$$\frac{R_m}{R_m+X}=\frac{D_m}{V}$$

or

$$R_m + X = \frac{VR_m}{D_m}$$

$$\therefore \qquad X = \frac{VR_m}{D_m} - R_m$$

in which R_m is resistance of voltmeter; D_m is deflection of voltmeter with resistance in series; V is voltage of the supply system when reading D_m; and X is resistance sought.

TYPICAL WIRING CONNECTIONS

DIRECT-CURRENT GENERATORS

Shunt Type, Supplying Constant-Potential Circuit. A shunt type d.c. generator is represented in Fig. 7, with the necessary connections. The brushes are connected to the two conductors forming the main circuit; also to the field-magnet coils *Sh* through a resistance-box *R*, to regulate the strength of the current and, therefore, the magnetism in the field. A voltmeter also is connected to the two brushes or main conductors to measure the voltage or electrical pressure between them. One of the main conductors is connected through an ammeter *A*, which measures the total current on the main circuit. The lamps *L*, or motors *M*, are connected in parallel between the main conductors or between branches from them. This represents the ordinary low-tension system for electric light and power distribution from isolated plants or central stations.

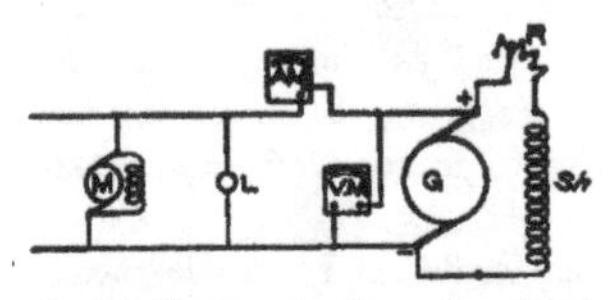

Fig. 7. Diagram for Connecting Shunt Type of D. C. Generator

Series Type. In a series-wound dynamo, the winding on the field being in series with the armature, the number of ampere turns in the field varies with the load on the machine or the amount of current flowing.

The field is wound with relatively few turns of large wire capable of carrying the full armature current. The voltage of a

INSTALLATION SHOWING GROUP DRIVE FOR MUSKEGAN MOTOR SPECIALTIES COMPANY. FAIRBANKS-MORSE INDUCTION MOTORS USED

Courtesy of Fairbanks, Morse and Company

series generator is low at light loads, but increases rapidly as load comes on. At normal load the field approaches the saturation point. Series generators are seldom, if ever, used except for constant-current generators on some arc-lighting systems or for boosters. Fig. 8 shows a series generator in which the load, field, and armature are all in series.

Compound Type. A compound-wound generator has two sets of field windings—a shunt winding and a series winding—and is therefore a combination of the two preceding types, Fig. 9. At no load and light load the shunt field Sh is excited to give normal voltage. The series winding S is connected to assist the shunt field so that, as the load comes on and a larger current passes through the series field, the drop in voltage which would otherwise occur is prevented. The voltage at no load is adjusted by means of the rheostat R in

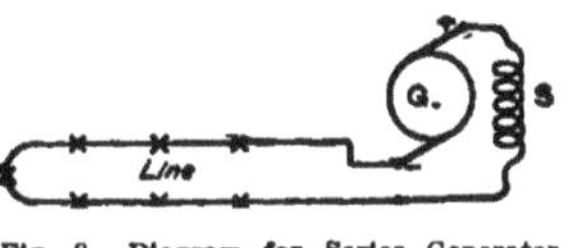

Fig. 8. Diagram for Series Generator Showing Load, Field, and Armature, All in Series

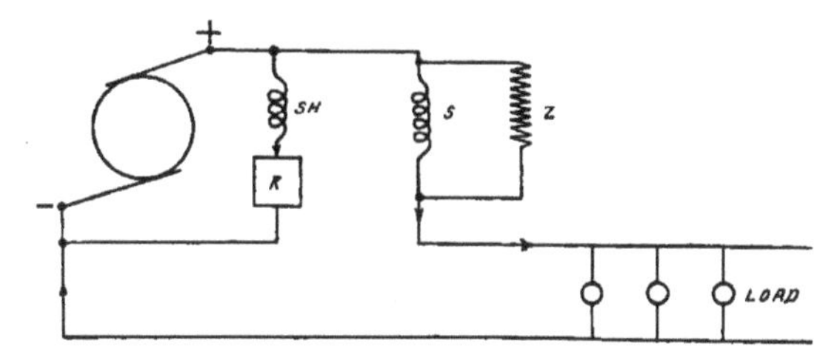

Fig. 9. Diagram for Compound-Wound Generator

series with the shunt field. As the load comes on, the voltage is automatically regulated by the action of the series field. The voltage may be held practically constant from no load to full load or it may be increased at full load to 10 per cent or 15 per cent above the no-load value. This is the usual practice in generators for railway work and is called "overcompounding."

The adjustment of the compounding is obtained by connecting a resistance Z, known as a series-field shunt, across the terminals of the series field. The relative value of the resistance of the series field and its shunt resistance determines the current in the series field, and by adjusting the shunt resistance any desired value of compounding may be obtained.

Direct-Current Generators in Parallel. Successful parallel operation of direct-current generators requires that corresponding polarities be connected together. The currents of the various generators will be added and may have different values, but the voltage will not be increased. At the time the machines are connected together, the voltages should be equal, to avoid any undue disturbances on the system.

Shunt-Wound Generators in Parallel. The connection of shunt-wound generators to switchboard buses is shown in Fig. 10.

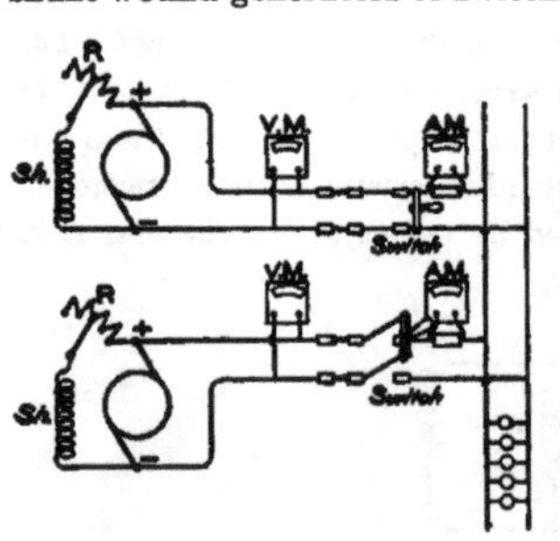

Fig. 10. Wiring Diagram for Two Shunt-Wound Generators in Parallel

The machine with its main switch closed is operating on the station bus, the other generator is brought up to exactly the same voltage and its main switch is closed. The voltage on this second machine is then raised by means of the shunt-field rheostat until the generator takes its proportion of the load. Assume the two generators to be of the same rating and voltage, and each, therefore, supplying one-half the total current. If, owing to a drop in speed or any other cause, the voltage of one generator drops, more load would be thrown on the other machine. As the tendency of a shunt generator is to increase its voltage as the load decreases and decrease its voltage as the load increases, the machine that has been relieved of its load would tend to raise its voltage and would immediately take more load. Shunt generators are, therefore, well adapted to parallel operation, but they are not often used because of poor voltage regulation.

Compound-Wound Generators in Parallel. A simple wiring diagram of two compound-wound generators connected in parallel is shown in Fig. 11. *A* is the armature; *B* is the series field; *C* is the shunt field; *D*, *E*, and *F* are single-pole single-throw switches; *G*, *H*, and *I* are buses; and *R* is the shunt-field rheostat. The switches *D* and *F* connect the main leads from the machine terminals to the main buses *G* and *I*. The switches *E* connect a point between the armature and the series field on each machine to an equalizing bus *H*.

Assume that No. 1 machine is in service and it is desired to

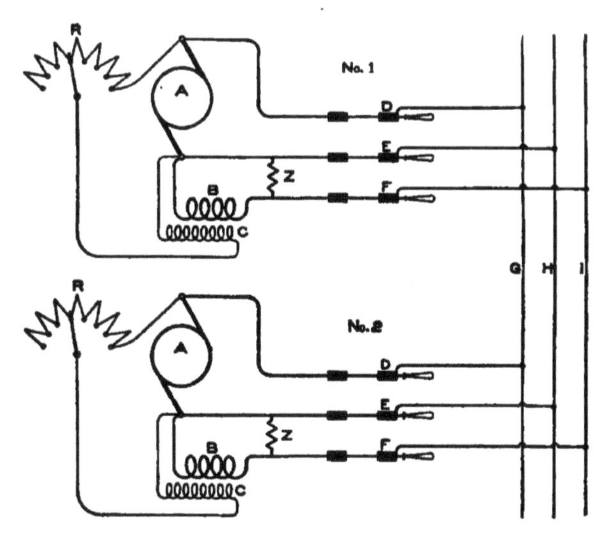

Fig. 11. Wiring Diagram for Two Compound-Wound Generators in Parallel

place No. 2 machine in parallel with it. Bring No. 2 machine up to speed and adjust its voltage to about normal. Close switch *F* and switch *E*. This places the series field of No. 2 in parallel with that of No. 1 and the voltage of No. 2 will be increased. Now adjust the voltage of No. 2 by means of the field rheostat until the voltage of both machines is the same. Next close switch *D*. The load on No. 2 can now be adjusted by means of the field rheostat until the two machines are dividing the load properly.

The voltage of the two machines can be compared in any one

of several ways. A voltmeter may be connected to the buses permanently and a second voltmeter arranged to be connected to any machine by means of plugs and receptacles, or a single voltmeter can be arranged to be plugged alternately to any two machines.

On small generators the three single-pole switches are often replaced by a triple-pole switch. In this case the voltages of the running and the incoming machines are adjusted to equal value and the switch on the incoming machine is closed. The resulting sudden shift of load may not be objectionable in small machines.

Before two machines can be paralleled, it must be determined whether the compounding is the same; that is, whether the rise in voltage from no load to full load is the same. A shunt resistance, Z, Fig. 11, is placed across the series field of each machine. By changing this resistance, the amount of current in the series field and thereby the amount of the compounding can be adjusted.

When two machines are operating in parallel and one machine supplies more than its share of current, the resistance of the path through its series field coil should be increased. Since the resistance of the series field and its connection to the switchboard is low, putting a longer lead between the series field and the switchboard will usually give the desired result.

Where very large generators are used, the cost of cables to the switchboard may be considerable. In such cases the usual practice is to place the equalizer bus under the machines and as close to them as possible. The equalizer switches are then placed on a base mounted on the frame of the machine, or on a pedestal mounted beside the machine.

Direct-Current Generators in Series. The arrangement of direct-current generators in series is not very common in ordinary power work. Constant-current series-wound generators have sometimes been connected in series for arc-light service. Probably the most common example of this practice has been the use of two shunt or compound-wound generators connected in series for Edison three-wire service. At present there are a large number of stations where shunt- or compound-wound machines are operated in series in order to obtain the high voltages now common in heavy traction service.

No special connection is necessary in the machine itself to

connect it in series with another. The negative of one machine is simply connected to the positive of the next one, thus adding the voltages.

Commutating-Pole Generators

Characteristics. In addition to the fundamental types of direct-current generators already described, there is the commu-

Fig. 12. Westinghouse Type Q Generator Showing Commutating Poles

tating-pole type of generator. This differs from the above, in mechanical respect, only in the fact that midway between each pair of main poles there is a small pole. The electrical connections differ from those of the non-commutating-pole machines only in the fact that the main current from the armature passes through the windings on the commutating poles, as well as, and in the same manner as, through the series fields already described for compound-wound generators. In the same manner the winding on the commutating poles may have a shunt across its terminals the same as the series field shunt Z shown in Fig. 9. The same remarks apply to commutating pole motors.

The object of the commutating poles is to prevent shifting of the non-sparking position of the brushes from no load to full load. This is accomplished, because the commutating pole produces the necessary flux for neutralizing the effect of armature reaction.

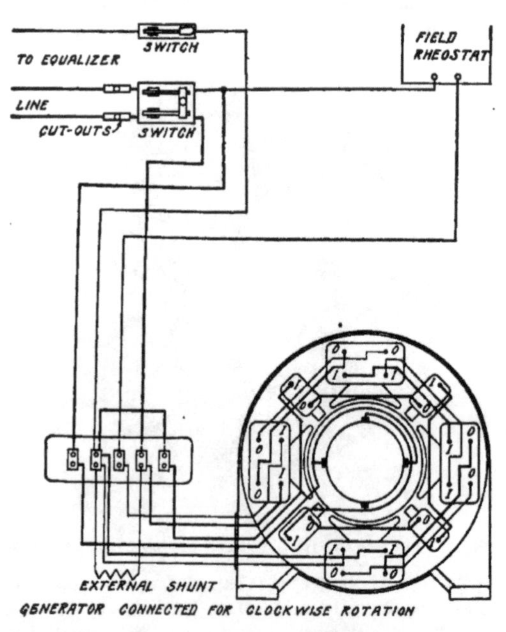

Fig. 13. Wiring Diagram for Commutating Four-Pole Compound Generator

Location of Brushes on Commutating-Pole Machines. Remove one brush from a brush holder and in its place insert a fiber brush of the same dimensions as the brush removed. Through the fiber brush bore two holes, about the size of a lead in a pencil, at such a distance apart that the holes are over the centers of two adjacent commutator bars. Insert in these holes the leads from two pencils so that the ends of the leads ride lightly on the commutator. To the other ends of the leads connect a low reading voltmeter or a millivoltmeter. The brushes should previously have been set in a line on the commutator passing through the center of the main pole piece. At no load excite the machine to give normal voltage and move the brush-holder yoke until the voltmeter connected to the exploring brush reads zero. This is the proper position for the brush and will be the same for full load, if the shunt around the commutating field has been properly adjusted.

Fig. 12 shows a commutating-pole machine. Both motors and generators have the same arrangement of poles. Fig. 13 shows the electrical connection of a four-pole commutating-pole generator.

SYNCHRONOUS CONVERTERS

Comparison with D.C. Generator. In general appearance and construction the synchronous converter resembles the direct-current generator. The chief difference in appearance is that the synchronous converter has a set of collector rings mounted on, and insulated from, the shaft usually on the opposite end from the commutator. Another difference is seen in the relative dimensions of the magnetic circuit, which is much smaller in the synchronous converter than in a direct-current generator of the same capacity. The commutator end of a synchronous converter, or rotary converter, as it is often called, is exactly like any direct-current generator. At the other end the collector rings are tapped to equidistant points of the armature winding. As many taps are taken from the armature winding to each collector ring as there are pairs of poles.

Uses. The rotary is commonly used for converting alternating current into direct current. The voltage of the applied alternating current bears a fixed and definite ratio to the value of the direct-current voltage on the commutator. However, the machine may be run as an inverted rotary, having direct current applied to the commutator and delivering alternating current at the collecting rings.

Voltage Relations in Converters. *Two-Ring Rotary.* To understand the voltage relations in converters, consider briefly the case of a two-ring (single-phase) rotary. The maximum value of alternating-current voltage will occur when the commutator bars connected through the armature winding to the collector ring pass under the brushes. Assume E_{dc} the value of direct-current voltage. This is equal to the maximum value V of the alternating voltage. The effective value* V_e is equal to the maximum value V divided by $\sqrt{2}$; that is V_e equals $\frac{V}{\sqrt{2}}$; since V equals E_{dc} then V_e equals $E_{dc} \times 0.707$ or V_e. That is, taking the voltage on the direct-current end of a single phase rotary to be 100 volts, the alternating-current voltage necessary to give this must be 70.7 volts across the rings. Let the effective value V_e for a single-phase rotary be represented by the diameter of a circle. Let Fig. 14 represent the relation of a. c. voltages in a three-ring three-phase rotary converter. *1*, *2*, and *3* represent the rotary rings where the impressed voltage is applied. V_s represents the value of voltage between any two rings. The following is a simple proof of the value of V_s in terms of V_e, and this in turn in terms of E_{dc}. Since we have already shown

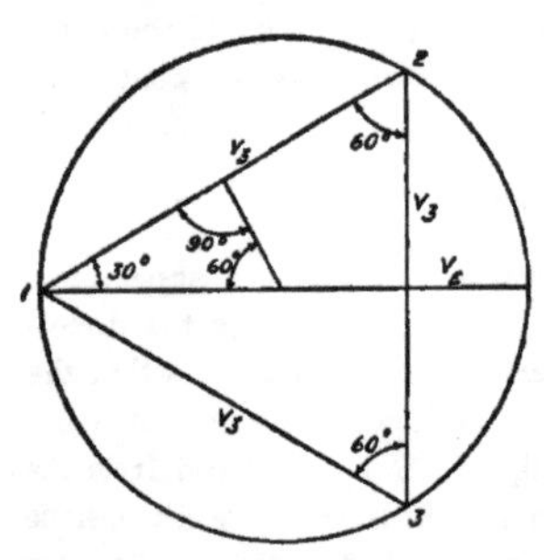

Fig. 14. Relation of A. C. Voltages in a Three-Ring Three-Phase Rotary Converter

$$V_e = \frac{E_{dc}}{\sqrt{2}}$$

$$\text{Cos } 30° = \frac{\frac{V_s}{2}}{\frac{V_e}{2}} = \frac{V_s}{V_e}$$

$$\text{Cos } 30° = \frac{1}{2}\sqrt{3} = \frac{V_s}{V_e}$$

* Alternating-current voltmeters read *effective* value of volts, not maximum value.

$$V_3 = \frac{\sqrt{3}}{2} \times V_e$$

But V_e was shown to equal $\frac{E_{dc}}{\sqrt{2}}$

$$\therefore \qquad V_3 = \frac{\sqrt{3}}{2} \times \frac{E_{dc}}{\sqrt{2}} = .612\, E_{dc}$$

Which means that the alternating-current voltage across any two collector rings is equal to the direct-current voltage (measured across brushes of opposite polarity) multiplied by 0.612. Conversely, the direct-current voltage is equal to the alternating-current voltage across any two rings in a three-ring (three-phase) rotary converter divided by 0.612.

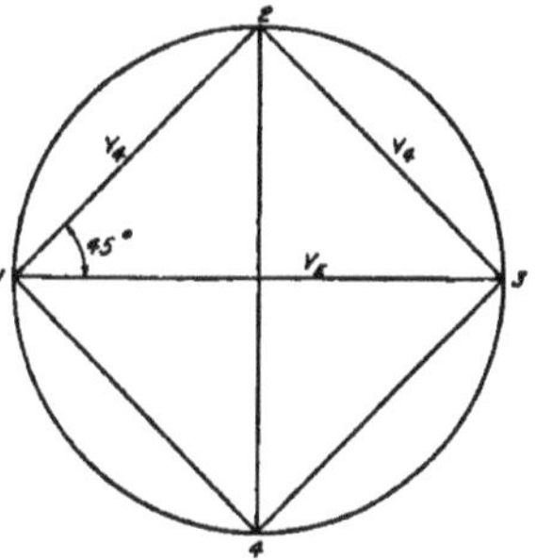

Fig. 15. Relation of A. C. Voltages in a Four-Ring, Two-Phase Rotary Converter

Four-Ring Rotary. Consider next the case of a four-ring (two-phase) rotary, and as in the previous case, let *1, 2, 3,* and *4* represent the rings. In Fig. 15

$$\text{Cos } 45° = \frac{\frac{V_e}{2}}{V_4} = \frac{V_e}{2V_4}$$

$$\text{Cos } 45° = \frac{1}{2}\sqrt{2} = \frac{V_e}{2V_4}$$

$$V_4 = \frac{V_e}{\sqrt{2}}$$

But $$V_e = \frac{E_{dc}}{\sqrt{2}}$$

$$\therefore \qquad V_4 = \frac{\frac{E_{dc}}{\sqrt{2}}}{\sqrt{2}} = .500\, E_{dc}$$

Consider then the voltages on a four-ring (two-phase) rotary in which *1, 2, 3,* and *4* are phases; the voltages between adjacent rings will be V_4 which equals .500 E_{dc}, and the voltages across each phase will be the same as for a single-phase rotary, that is,

$$V_e = .707\, E_{dc}$$

Six-Phase Rotary. Fig. 16 represents a six-phase relationship of voltages. The collector rings are numbered *1, 2, 3, 4, 5, 6.* Consider the small triangle *1-2-a.*

$$\text{Sin } 30° = \frac{\frac{V_e}{4}}{V_6}$$

$$\text{Sin } 30° = \frac{1}{2}\sqrt{1} = \frac{V_e}{4V_6}$$

$$V_6 = \tfrac{1}{2} V_e$$

But $$V_e = \frac{E_{dc}}{\sqrt{2}}$$

$\therefore$ $$V_6 = \frac{1}{2} \times \frac{E_{dc}}{\sqrt{2}} = .354\, E_{dc}$$

Note that Fig. 16 shows single-phase, three-phase, and six-phase relationships of voltages. V_6 is the voltage between any two adjacent rings represented by the sides of a hexagon inscribed in a circle whose diameter is equal to the effective voltage across the rings of a single-phase rotary. Measured in terms of the d. c. voltage, we have shown that V_6 equals $E_{dc} \times .354$.

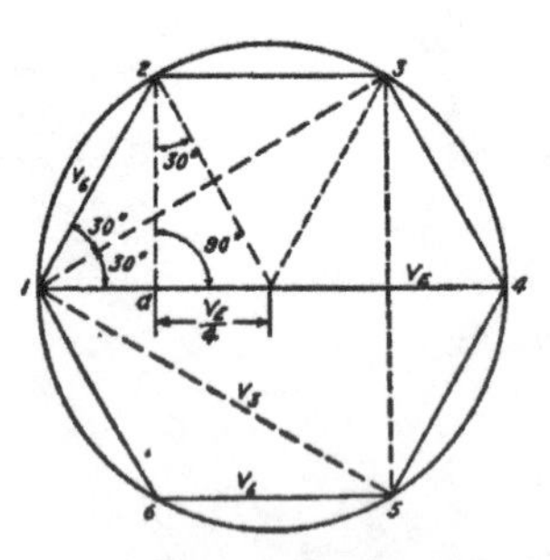

Fig. 16. Relation of A. C. Voltages in a Six-Phase Rotary Converter

In the dotted triangle *1-3-5* it will be noted the lines *1-3, 3-5,* or *5-1* represent the voltage between collector rings, that is V_3. We have already shown that V_3 equals $E_{dc} \times .612$.

A six-phase rotary has voltages impressed directly upon collector rings *1* and *4, 2* and *5,* and *3* and *6,* which, from Fig. 16, will be seen to be equal to diameters of the circle. Therefore the same relation exists as for a single-phase rotary. Thus the alternating-current voltage across rings *1* and *4*, *2* and *5*, and *3* and *6* is equal to d. c. voltage multiplied by .707.

From the above explanation, and referring to Fig. 16, it will be seen that

a. c. volts across *1* and *2* $= E_{dc} \times .354$
a. c. volts across *1* and *3* $= E_{dc} \times .612$
a. c. volts across *1* and *4* $= E_{dc} \times .707$

Fig. 17. General Electric 600-Volt Compound-Wound Rotary Converter

Conversely, $E_{dc} = V_6$ between rings *1* and *2* $\div .354$
$E_{dc} = V_3$ between rings *1* and *3* $\div .612$
$E_{dc} = V_e$ between rings *1* and *4* $\div .707$

A modern type of railway rotary converter of the compound-wound type is shown in Fig. 17.

Switching of Synchronous Converters. The following instructions apply to the usual connections and to the switchboard equipment of General Electric compound-wound converters in railway service. Referring to Figs. 18 and 19, a machine should be started as follows, it being assumed that all switches are open before starting and that the potential plug is inserted in the potential receptacle on the direct-current switchboard panel, thus connecting the machine to the voltmeter. (See note.)

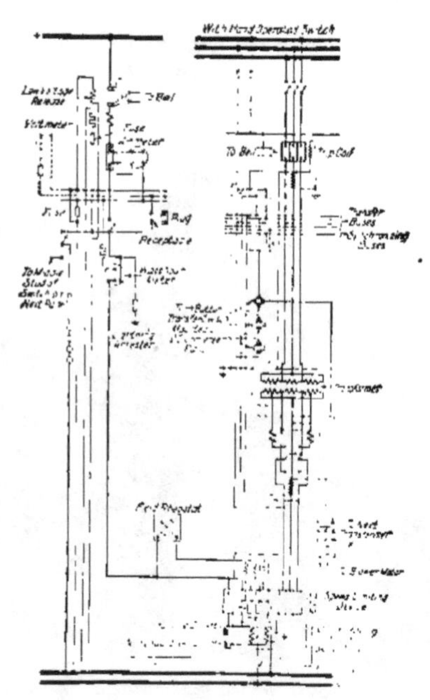

Fig. 18. Switching Diagram for Three-Phase Converters
Courtesy of General Electric Company

First: Close main high-tension switch *A*.

Second: Close starting switch *B* upward. (For six-phase machines with tandem switches, both switches *B* and *B B* should be closed upward.)

The machine will then run up to speed and lock in step, which will be indicated by a cessation of the beats of the direct-current voltmeter.

Third: Close equalizer switch *G*.

Fourth: Close field break-up and reversing switch *E* into the top position.

Fifth: For three-phase machines throw starting switch *B* quickly from top to bottom contacts, Fig. 18. With six-phase machine, Fig. 19, throw starting switch *B B* quickly from top to

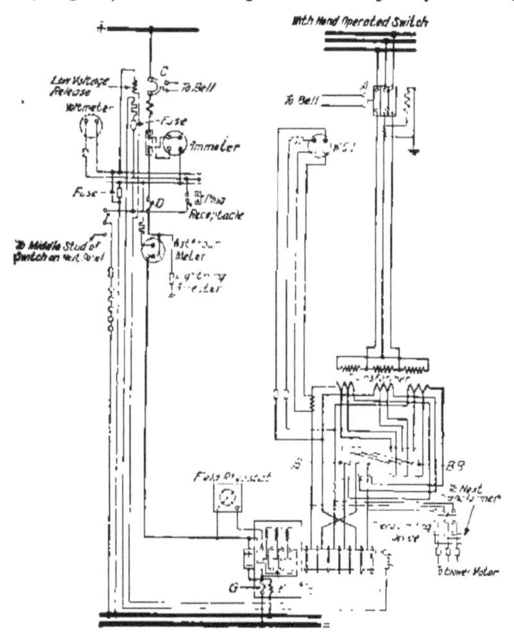

Fig. 19. Connections of Starting Switches for Six-Phase Converters
Courtesy of General Electric Company

bottom contacts and then switch *B* from top to bottom contacts. Finally open switch *B B*.

NOTE: If other machine or machines are carrying load when a compound-wound converter is started, the correct polarity may be insured by closing the equalizer switch *G* when the machine locks in step. By watching the swings of the direct-current voltmeter as the machine approaches synchronism, switch *G* may be closed just previous to the last two or three swings, thus insuring proper locking on the first trial, if there is current for the series field from other machines. If the machine locks with the wrong polarity as

indicated by the direct-current voltmeter needle going down off the scale, the field switch *E* must be closed first into the down position, which will cause the voltmeter to return above zero, when the switch *E* must be pulled out and closed into the top position. The reversal of polarity should be made while the machine is running on first starting tap.

Sixth: Adjust direct-current voltage to approximately that of the busbars.

Seventh: Push up the low voltage release of circuit breaker and close circuit breaker *C*.

Eighth: Close main switch *D*.

Ninth: Adjust the division of loads between machines, if more than one are in service, by means of field rheostats.

NOTE: Machines as now built do not have a shunt to the series field, so that the series field shunt switch exists only on older machines.

Shutting Down Synchronous Converter. To shut down a synchronous converter, open the circuit breaker *C*, pull out and turn the circuit-closing auxiliary switch to stop the ringing of the alarm bell; open the main switch *D;* open the high-tension alternating-current oil switch *A;* allow the machine to run down in speed until the voltage falls off to about one-fifth normal value before opening the field switch *E* or the starting switch *B*; open the field switch *E*, equalizer switch *G*, and starting switch *B*.

THREE-WIRE SYSTEM

Two Direct-Current Generators on Three-Wire System. In the ordinary three-wire system for incandescent lighting and power service, no particular precautions are required in starting or connecting the machines; and either of the arrangements shown in Figs. 20 and 21 may be adopted. The two sides of the system are almost independent of each other, and form practically separate circuits, for which the middle or neutral wire acts as a common conductor. There is, however, a tendency for the dynamos, Fig. 20, to be reversed in starting up, in shutting down, or in the case of a severe short circuit. This

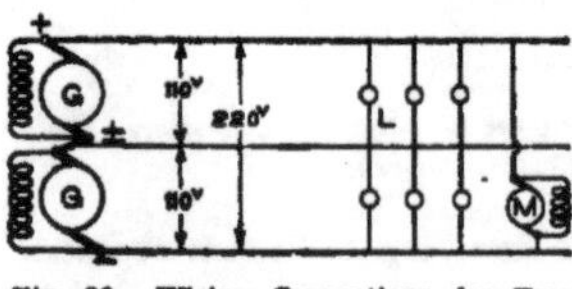

Fig. 20. Wiring Connections for Two D. C. Generators on Three-Wire System

can be avoided by exciting the field coils of all the dynamos from one side of the system, or from a separate source. To obtain good regulation, it is necessary to balance the load equally on both sides of the system. It is advisable to employ 220-volt motors on 110-volt 3-wire systems, and to connect them across the outside conductors so that the motor load shall not unbalance the system.

Generators with Balancer Set. Fig. 21 represents a 220-volt generator connected to the outside wires of a three-wire system. Two 110-volt machines are connected in series across the system, and the neutral conductor connects with the intermediate point between them. If the system is perfectly balanced, both 110-volt machines operate as motors without load. As the system becomes unbalanced, the machine on the heavier loaded side begins to operate as a generator and the other as a motor. The machines must be mechanically connected but can be separately regulated to give any desired voltage division between sides, and each machine can be compounded to compensate for unequal line losses and neutral drop when the system is unbalanced.

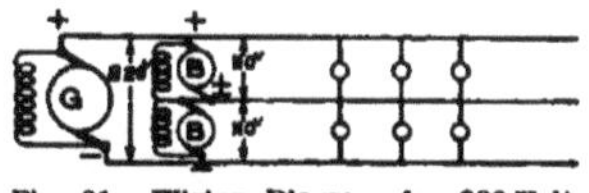

Fig. 21. Wiring Diagram for 220-Volt Generator with Balancer Set on Three-Wire System

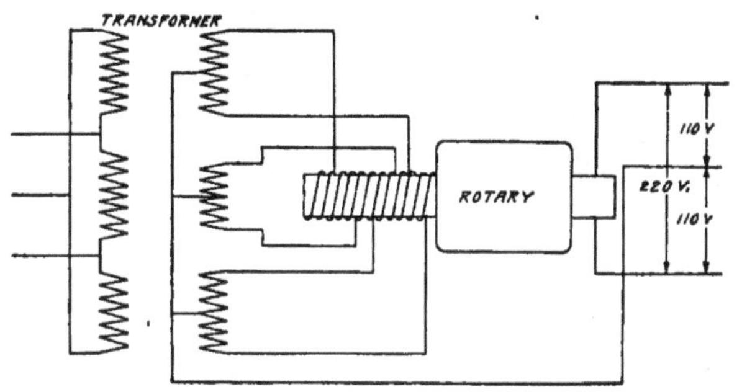

Fig. 22. Wiring Diagram for Three-Wire System with Rotary Converter Showing Transformer Connections

Three-Wire Systems from Rotary Converters. Where rotary converters are used to supply a three-wire system, the neutral conductor can be connected either to the middle points of the transformer secondaries, or to a compensator connected to the alternating-current leads.

The arrangement of the transformers or compensator windings must be such that they cannot operate reactively to prevent the current flow necessary to maintain the neutral, Fig. 22. Three-wire generators are often used for supplying three-wire systems. This machine is like a two-wire generator but has the armature winding tapped at suitable points. These taps are connected to a

Fig. 23. Crocker-Wheeler 75 Kw. Three-Wire Generator

compensator or a balance coil from the middle point of which the neutral is obtained. Several forms of this type have been used by various manufacturers. One of these has the armature taps brought out to collector rings. The compensator is then mounted separately from the machine and connected through brushes to the rings.

Since this form requires renewals and attendance on two or more sets of brushes and rings, some manufacturers have provided

a winding in the armature to generate the voltage supplying the neutral conductor, which thus requires but one set of brushes and one collector ring, Fig. 23, in addition to the commutator brushes.

A later type of machine has a revolving compensator. This consists of a circular magnetic core upon which are mounted the coils. The core and coils are bolted directly to the back end of the armature spider. The single neutral connection is carried through the spider to a single collector ring mounted on the outer end of the commutator shell. This form has the advantages of few collector rings and brushes, small space, and greater simplicity in wiring.

DIRECT-CURRENT CONSTANT-POTENTIAL MOTORS

Shunt-Wound Motors. A motor to operate at nearly constant speed, with varying loads, on a d.c. constant-potential system (110- or 220-volt lighting circuits) is usually plain shunt-wound. This is the commonest form of stationary motor. The field coils are wound with wire of such a size as to have the proper resistance and resulting magnetizing current; and since the potential applied is practically constant, the field strength is constant.

Starting. In starting shunt motors, no trouble is likely to occur in connecting the field to the circuit. The difficulty is with the armature current, because the resistance of the armature is very low in order to get high efficiency and constancy of speed, and the rush of current through it in starting might be twenty or more times the normal number of amperes. To avoid this excessive current, motors are started on constant-potential circuits through a rheostat or starting box containing resistance coils.

The main wires are connected through a branch cut-out (with safety fuses), and preferably also a double-pole knife switch *Q*, to the motor and box, as indicated in Fig. 24. When the switch *Q* is closed, the arm *S* being in its left-hand position, the field circuit is closed through the contact stud *f*, and the armature circuit is closed through the resistance coils *a a a* which prevent the rush of current referred to. The motor then starts and, as the speed rises, it generates a counter e.m.f., so that the arm *S* can be turned as shown until all the resistance coils *a a a* are cut out.

The arm S should positively close the field circuit first, so that the magnetism reaches its full strength (which may take several seconds) before the armature is connected.

In the arrangement shown in Fig. 24, the release magnet has its coils in series with the field. As long as the motor is in operation, the core is energized and the arm S is held in the position shown. If, however, the current applied to the motor is cut off and the motor comes to rest, the core of the magnet loses its attrac-

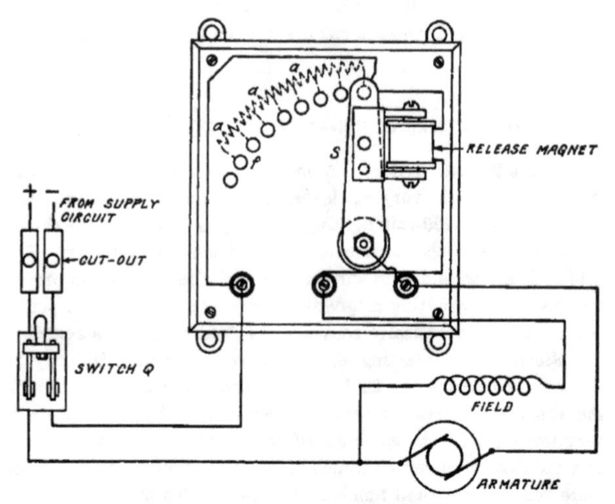

Fig. 24. Wiring Diagram for Connections of Shunt-Wound Motor to Starting Box

tive force, and the arm S is released, being automatically moved back to the starting position by a spring.

The coils $a\,a\,a$ are made of comparatively fine wire which can carry the current only for a few seconds in a starting box; but if the wire is large enough to carry the full current continuously, it is called a "regulator", because the arm S may be left so that some of the resistances $a\,a\,a$ remain in circuit, and they will have the effect of reducing the speed of the motor, which is often very desirable.

In some cases where a circuit is used exclusively for a single motor, the speed is regulated without heavy resistances by varying the e. m. f. of the dynamo which supplies the circuit. The dynamo regulator is then placed near the motor. The advantage is that the regulator is not compelled to control a heavy current, but a special circuit of unvaried pressure must be provided to keep the field of the motor constant.

Speed Control. The speed control of a shunt motor may be simply obtained as follows:

Fig. 25. Fort Wayne Northern Type B Shunt-Wound Motor

(a) For lower speeds, insert resistance in series with the armature circuit. The resulting i. r. drop reduces the value of the voltage applied to the armature terminals, and thus reduces the speed.

(b) For higher speeds, insert resistance in the shunt-field circuit. This reduces the magnetic flux and, to generate the same counter e. m. f., the motor must speed up.

The field circuit of a shunt motor should never be opened while pressure is still applied to the armature terminals, as under these conditions the armature current becomes very excessive and the armature is likely to race and be damaged. A moderate decrease in field strength only is allowable; otherwise sparking becomes excessive. Fig. 25 shows a Fort Wayne Northern Type B shunt-wound motor.

Series-Wound Motors. The ordinary electric railway motor on the 600-volt system is the chief example of the class, Fig. 26. Motors for fans, pumps, or electric elevators and hoists are either of this kind or of the compound type. A rush of current tends to occur, when the series type of motor is started, similar to that in the case just described; but it is less, because the field coils are in series, so that their resistance and self-induction reduce the excess.

Fig. 26. Box Frame Railway Motor
Courtesy of General Electric Company

Furthermore, the counter e.m.f. is greater even at low speed because the heavy current produces a strong field.

The connections, as indicated in Fig. 27, are very simple, the armature, field-coils, and rheostat all being in series and carrying the same current.

The series-wound motor on a constant-potential circuit does not have a constant field strength, and does not tend to run at constant speed, like a shunt motor. In fact it may "race" and tear itself apart if the load is taken off entirely; it is, therefore, suited only to railway, pump, fan, or other work where variable speed is desired, and where there is no danger of the load being removed or a belt slipping off. It is also used where the potential is subject to sudden and large drops, as on the ends of long trolley circuits, because in such a case a shunt motor becomes momentarily a generator and sparks very badly. The fields of series motors are sometimes "overwound," that is, so wound that they will have their full strength with even one-half or one-third of the normal

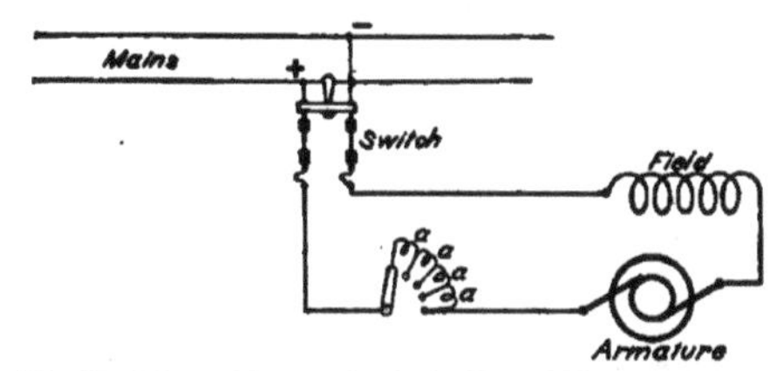

Fig. 27. Wiring Diagram for Series-Wound Motor Connections

current. The objects are: to secure a nearly constant speed with varying loads; to enable the motor to run at high efficiency when drawing small currents; and to prevent sparking at heavy loads.

In multipolar motors having more than two field-coils, the coils are all connected together, and are equivalent to the single

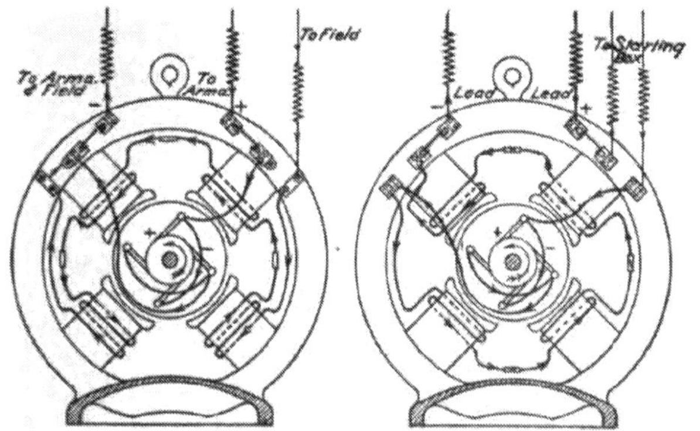

Fig. 28. Wiring Diagram for Four-Pole Shunt and Series-Wound Motors

pairs of coils shown in the several diagrams. Being separated, however, it is sometimes necessary to trace out the connections. Fig. 28 represents the necessary connections for a four-pole motor, shunt-wound and series-wound.

Differentially-Wound Motors. This is a shunt-wound motor with the addition of a coil of large wire on the field, connected in series with the armature in such a way as to oppose the magnetizing effect of the shunt winding and weaken the field, thus causing the motor to speed up when the load is increased, as an offset to the slowing-down effect of load.

It was formerly used for obtaining very constant speed, but it has been found that a plain shunt motor is sufficiently constant for almost all cases. The differential motor, if overloaded, has the great disadvantage that the current in the opposing (series) field-

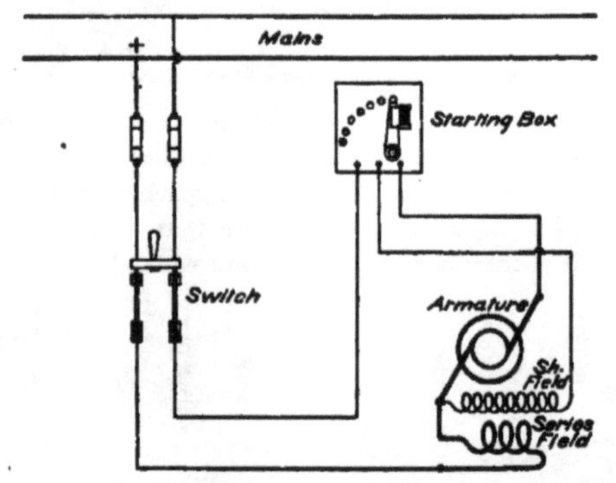

Fig. 29. Wiring Diagram for Compound-Wound Motor Connections

coil becomes so great as to kill the field magnetism; and instead of increasing or keeping up its speed, the armature slows down or stops and is likely to burn out; whereas a plain shunt motor can increase its power greatly for a minute or so when overloaded, and will probably throw off the belt or carry the load until the latter decreases to the normal amount.

Compound-Wound Motors. This type of motor is also provided with both shunt- and series-field windings, Fig. 29, but in this instance they magnetize the field in the same direction, or, in other words, their effect is cumulative. This type of motor possesses the powerful starting torque feature of the series motor, but a less variable speed with varying loads. It is employed where a great starting torque and a fairly uniform running speed are required, as, for example, with electric hoists or elevators.

Method of Reversing Direction of Rotation in Motors. To reverse the direction of any direct-current motor, the direction of the current in either the field or the armature must be reversed. The direction of rotation would not be changed if the current in both the field and the armature were reversed. If the motor is

Fig. 30. Frame and Field Coils of Commutating-Pole Motor
Courtesy of Crocker-Wheeler Company

compound or differentially wound and the direction of rotation is to be reversed, the current in the series winding must be kept so that it either assists or opposes, as the case may be, the current in the shunt field, whether the rotation is reversed by changing the current through the armature or the shunt field.

Commutating-Pole Motors. The general remarks as to mechanical and electrical points of difference between commutating-

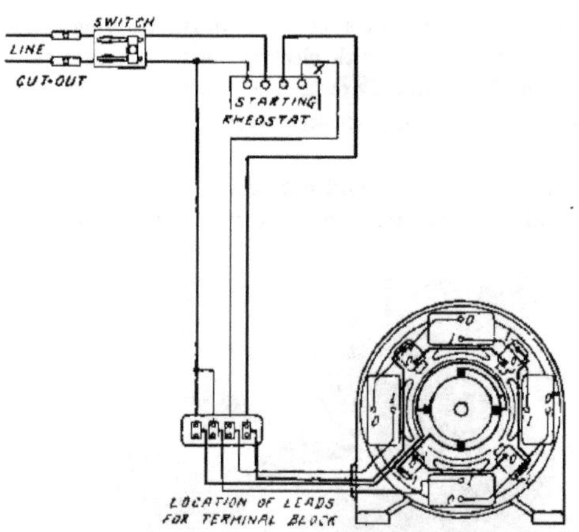

MOTOR CONNECTED FOR COUNTER CLOCKWISE ROTATION

31. Wiring Diagram of Rotation Connections for Commutating Four-Pole Shunt Motors

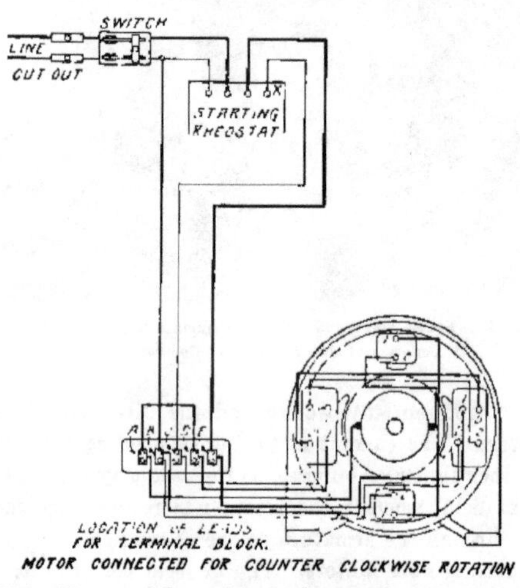

MOTOR CONNECTED FOR COUNTER CLOCKWISE ROTATION

32. Wiring Diagram of Connections for Commutating-Pole Bipolar Compound Motors

and non-commutating-pole types of generators apply to similar types of motors.

The electrical neutral on commutating-pole motors is determined by shifting the brushes until the same speed is obtained in both directions with the same value of field current and the same voltage on the d. c. mains.

Fig. 30 represents the frame and field coils of a standard commutating-pole motor, and Figs. 31 and 32 show the method of wiring commutating-pole motors for both the four-pole shunt type and two-pole compound type.

Fig. 33. Westinghouse Type E, 75 K. V. A. Generator, Showing Collector End

ALTERNATING-CURRENT GENERATORS

The present-day alternator consists of a stationary armature, called the stator, wound to generate either single or polyphase electromotive force, and a revolving field, called the rotor, excited from a separate 125- or 250-volt source. The exciting current is brought to the revolving field through collector rings mounted on and insulated from the shaft, the ends of the field winding being connected to these rings. Fig. 33 shows such a machine designed

Fig. 34. Three-Phase Crocker-Wheeler Alternator, Showing Separate D.C. Exciter on the Same Shaft

for direct connection to its prime mover. Machines of this class are quite often built with a small direct-current exciting generator mounted on the same shaft, as shown in Fig. 34. Small alternators

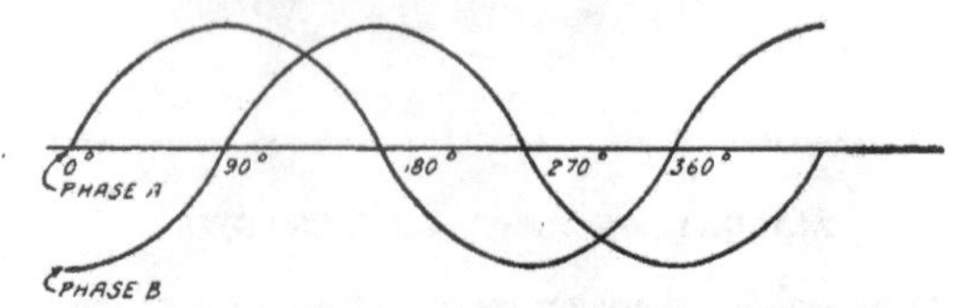

Fig. 35. Voltage Curves for Two-Phase Alternator

sometimes have revolving armatures, the current, single or polyphase, being taken from collector rings connected to the armature windings.

Single-Phase Type. There are a number of plants in this country still operating single-phase systems, but the advantages of using polyphase motors rather than single-phase motors in most power installations has caused this system to be practically abandoned. The single-phase generator need not be considered in discussing modern practice, since it is so little used.

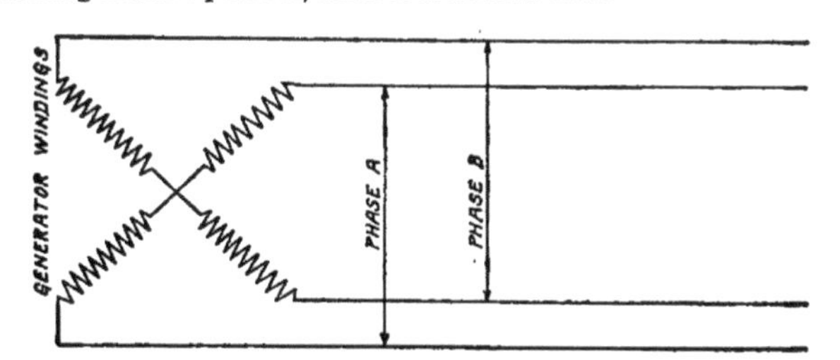

Fig. 36. Wiring Diagram for Two-Phase Four-Wire Generator with Armature Windings Ninety Degrees Apart

Two-Phase Type. A two-phase alternator has two armature windings arranged so that their voltages are 90 degrees out of phase. Fig. 35 shows the two voltage waves plotted in their relations, one being one-fourth of a cycle ahead of the other. Fig. 36 shows a two-phase four-wire generator with the armature wind-

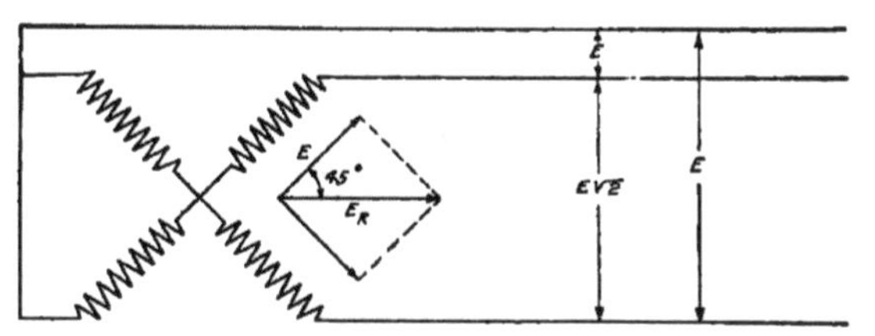

Fig. 37. Wiring Diagram for Two-Phase Generator Arranged for Three-Wire System

ings represented 90 degrees apart. The voltages on phase A and phase B are equal.

Two-phase generators are sometimes arranged with their two sets of armature windings connected together, as shown in Fig. 37. In this case only three wires are carried out to the distributing system. If the voltage of either phase is equal to E, then the voltage across the free ends of the inter-connected phase will be equal to $E \times \sqrt{2}$. (See Fig. 37.)

$$\text{Cos } 45° = \frac{\frac{1}{2}E_R}{E}$$

$$\text{Cos } 45° = \frac{1}{2}\sqrt{2}$$

$$\frac{1}{2}\sqrt{2} = \frac{\frac{1}{2}E_R}{E}$$

$$E_R = E \times \sqrt{2}$$

The current in any wire of a two-phase four-wire system is

$$I = \frac{KW}{2E}$$

in which KW is capacity of generator; E is voltage of phase A or B.

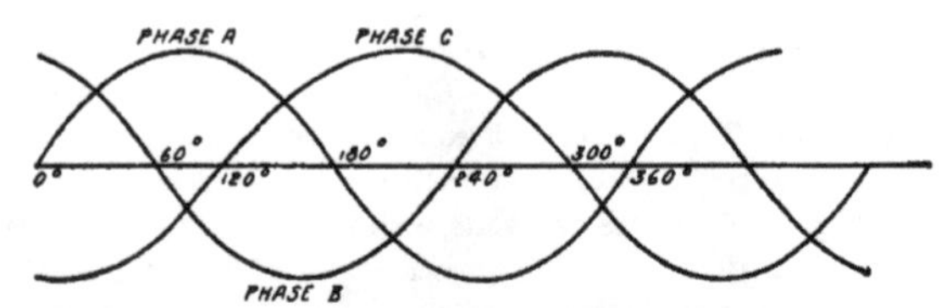

Fig. 38. Voltage Curves for Three-Phase Generator

Three-Phase Type. Three-phase generators have their armature windings divided into three sets of coils so arranged as to produce electromotive forces 120 degrees apart, Fig. 38. Armatures may be either Y- or Δ-connected, Figs. 39 and 40. In a Y-connected system, the line voltage is the phase voltage multiplied by $\sqrt{3}$ as shown, thus:

$$\text{Cos } 30° = \frac{\frac{1}{2}E_R}{E}$$

$$\text{Cos } 30° = \frac{1}{2}\sqrt{3}$$

$$\frac{1}{2}\sqrt{3} = \frac{E_R}{2E}$$

$$\therefore \quad E_R = E\sqrt{3}$$

the line current is found to be the same as the current per phase. In a Δ-connected system, the line voltage is the same as the phase voltage, but the current per line is the $\sqrt{3}$ times the current per phase. In either system

$$W = EI\sqrt{3}$$

for noninductive circuit, in which W is watts output; E is pressure in volts, and I is current in amperes.

For an inductive circuit whose power factor is less than unity,

$$W = EI\sqrt{3}\cos\phi$$

in which $\cos\phi$ is power factor.

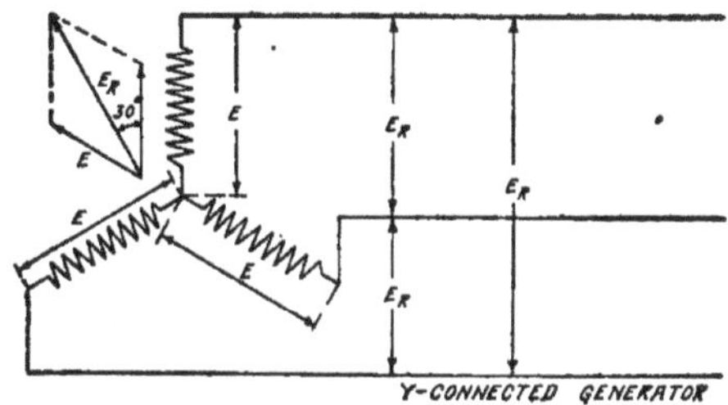

Fig. 39. Wiring Diagram for Y-Connected Three-Phase Generator

Alternators in Parallel. To run two alternators in parallel, several conditions have to be fulfilled: The incoming machine—as in the case of direct-current machines—must be brought up to

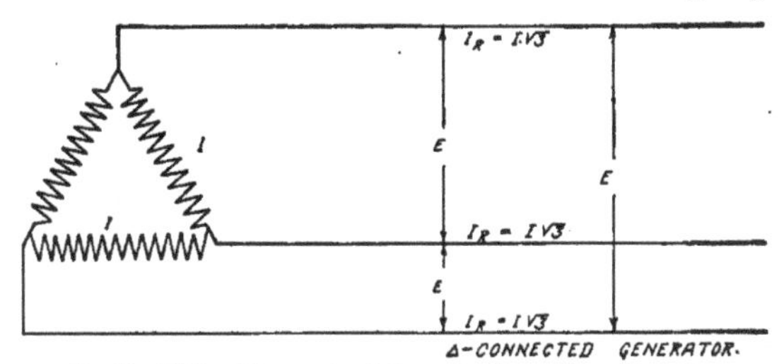

Fig. 40. Wiring Diagram for Δ-Connected Three-Phase Generator

nearly the same voltage as the first one; it must operate at exactly the same frequency; and, at the moment of switching in parallel, it must be in phase with the first machine. This correspondence of frequency and phase is called "synchronism."

Synchronizing Alternators by Lamp Indicator. It is impossible with mechanical speed-measuring instruments to determine the speed as accurately as is necessary for this purpose. There is, however, a very simple method of electrically determining small differences in speed or frequency. In Fig. 41, let *M* and *N* represent two single-phase alternators, which can be connected by means of the single-pole switch *A B*. Across the terminals of the switch is connected an incandescent lamp *L*, capable of standing twice the voltage of either machine. When *A B* is open, the circuit between the machines is completed through *L*. The two machines may be connected in parallel as follows: Assume machine *M* already in operation; bring up machine *N* to approximately the proper speed and voltage; then watch lamp *L*. If machine *N* is running a very little slower or faster than machine *M*, the lamp *L* will glow for one

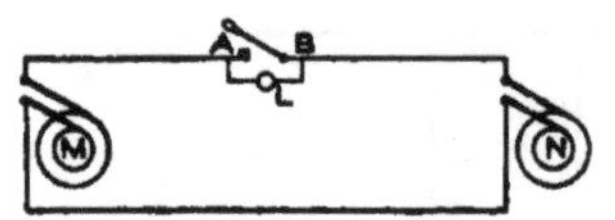

Fig. 41. Diagram of Two Single-Phase Alternators Arranged in Parallel

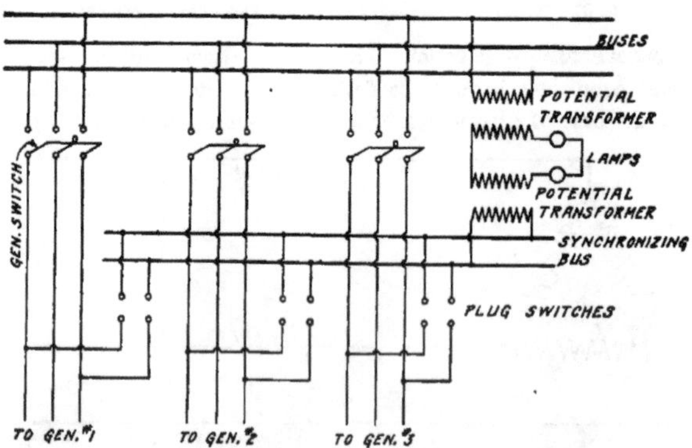

Fig. 42. Diagram of Connections for Synchronizing High-Voltage Alternators Through Step-Down Transformers

moment and be dark the next. At the instant when the voltages are equal in pressure and phase, *L* will remain dark; but when the phases are displaced by half a period, the lamp will glow at its maximum brilliancy. Since the flickering of the lamp is dependent upon the difference in frequency, the machines should not be thrown in parallel while this flickering exists. The prime mover of the incoming machine must be brought to the proper speed; and the nearer machine N approaches synchronism, the slower the flickering. When it is very slow, and at the instant when the lamp is dark, throw the machine in parallel by closing the switch across $A\,B$. The machines are then in phase, and tend to remain so, since if one slows down, the other will drive it as a motor. It is better to close the switch when the machines are approaching synchronism rather than when they are receding from it; that is, at the instant the lamp becomes dark.

Fig. 42 shows the method of synchronizing high-voltage alternators through step-down transformers. When two three-phase alternators are first placed in operation, synchronizing connections should be made across each phase. If all the lamps become bright or dark simultaneously, the alternators are ready for parallel operation. After all phases have once been tested, it is only necessary to compare a corresponding phase from each machine to indicate synchronism.

The connections, as shown in Fig. 42, indicate synchronism when the lamps are dark. If it is desired that a condition of synchronism shall be indicated when the lights are at maximum brightness, reverse the secondary connection of either one of the potential transformers.

Fig. 43. Weston Synchroscope Showing Internal Construction

Synchronizing Alternators by Synchronizer. Synchronizing by means of lamps is not considered accurate enough in the modern station. There is manufactured an instrument called a "synchronizer," or "synchronism indicator," which gives much more accurate results. One form of this instrument, shown in Fig. 43, has a revolving pointer which indicates, *by its position,* the exact point of synchronism; *and by its direction of rotation,* whether the incoming machine is running too slow or too fast. Lamps are usually provided with this instrument as a check on its operation and for use in case of a failure of the instrument due to a burn-out or other cause. In the type shown, the lamp is back of the scale and lights up only when the machines are at or near synchronism.

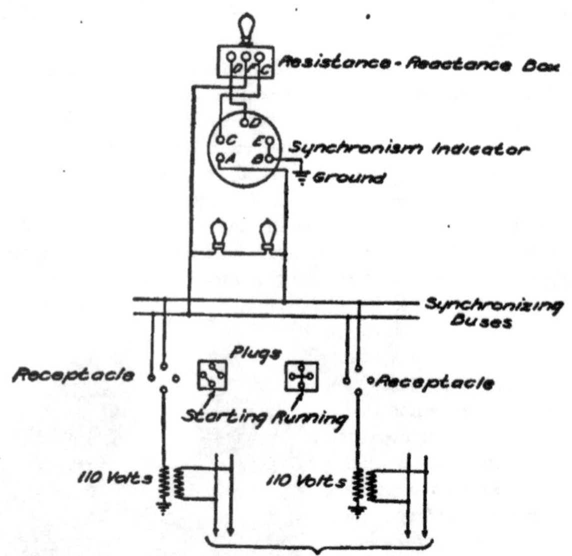

Fig. 44. Wiring Diagram for Synchronism Indicator with Potential Transformers, Secondaries Grounded. Synchronism Is Indicated When the Lamps Are Dark

Figs. 44 and 45 show two methods of connecting a synchronizer to two machines (one running and one starting) by means of plugs and receptacles mounted on the switchboard. Where the secondaries of the potential transformers can be grounded, which is usually the case, a common ground return can be used and the connections are much simplified.

When two alternating-current machines have been connected

INTERIOR OF GENERATING STATION OF THE ONTARIO POWER COMPANY, NIAGARA FALLS, ONT.

The generators are direct-connected to horizontal double turbines.

in parallel, the division of load should be adjusted. This cannot be accomplished as in direct-current machines by adjustment of the field rheostat. Change in field strength will cause more current to flow, but it will be 90 degrees out of phase with the potential and will not represent actual power. The only way to make the two machines supply proportional amounts of power is to adjust the input to their respective prime movers. The governor of the

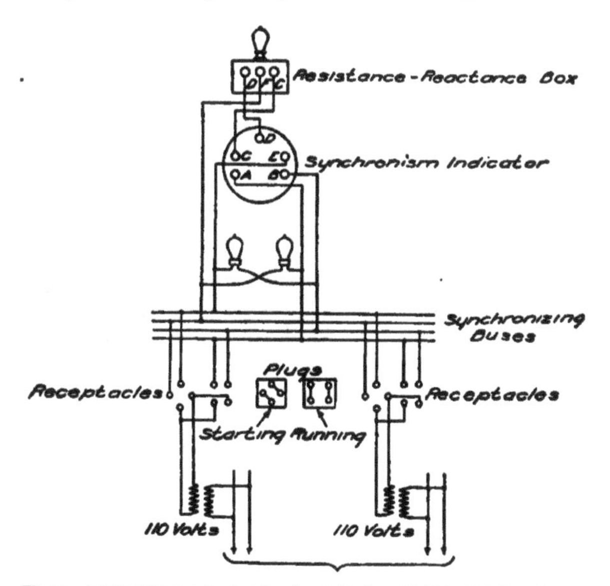

Fig. 45. Wiring Diagram for Synchronism Indicator with Potential Transformers, Secondaries Ungrounded. Synchronism Is Indicated When the Lamps Are Dark

prime mover should be adjusted as if for an increase in speed to make the machine carry more load, and conversely to make the machine carry less load.

If it is found necessary to increase the voltage of machines operating in parallel, the rheostats of all machines should be adjusted proportionally. If the rheostat of only one machine is shifted, cross currents will be caused to flow between machines, which do not represent actual power but do cause undesirable heating of the machines.

ALTERNATING-CURRENT MOTORS

Alternating-current motors may be generally classed as induction and synchronous types. The induction type of motor is particularly rugged in design, is simple of operation, requires little attention and a minimum of accessories; it does not tend to disturb the wave form of the system on which it is operating, and with normal load its power factor is fixed.

The synchronous motor closely resembles the revolving-field type of alternator and, in general, any standard type of alternator may be operated as a synchronous motor. This type of motor, therefore, requires practically the same number of accessories as an alternator. It requires care and skill on the part of the operator both in starting and in making the adjustments necessary for proper operation on the system. Since it is a synchronous piece of apparatus, like the generators supplying current to it, it may cause disturbance on the system on which it is operating unless very carefully handled. The power factor of this type of motor may be controlled by varying the excitation of its field. This feature allows the synchronous motor to be used for correcting the power factor of a system and causes its use in a large number of cases where the induction motor would otherwise be used because of its greater simplicity and ease of operation.

Induction Motors

Squirrel-Cage Type. The simplest form of polyphase induction motor is the "squirrel-cage" type, so-called because the rotating element with its windings short-circuited by a heavy copper end ring resembles a squirrel cage. Fig. 46 shows the rotor from a machine of this type. This rotor, also known as the secondary or armature, is made up of thin punched discs of specially treated iron. These punchings are provided with slots in which the windings are placed.

The stator, called also the field or primary, has windings distributed in the same manner as the winding in an alternator armature. The ends of the windings are brought out to a terminal board mounted on the frame of the motor, and the service lines are connected to the motor terminals. No connection is made to the rotor from any external source. Fig. 47 shows an exploded view of a motor of the type described.

The polyphase currents in the stator windings produce a revolving magnetic field. This in turn induces currents in the short-circuited winding on the rotor, which, on account of the reaction with the revolving magnetic field, causes the rotor to revolve. In order for any current to be induced in the windings of the rotor there must be a difference in speed between the revolving magnetic field and the rotor. It therefore follows that an

Fig. 46. Showing End Construction of Squirrel-Cage Rotor
Courtesy of General Electric Company

induction-motor rotor never can revolve at exactly the same speed as the magnetic field of the stator; that is to say, it cannot attain synchronous speed. The difference between the speed of the rotor and the revolving magnetic field is called the "slip" of the motor. In motors of this class the slip varies from 2 to 6 per cent, depending on the size of the motor, large motors having the lower percentage of slip. Motors of the squirrel-cage type take a current equal to 3 to 4½ times normal full-load running current to produce full-load starting torque. The normal full-load efficiency of two- or three-phase motors of this type varies approximately from 87 to 93 per cent, depending on the size of the motor. The power factor at full load varies from 85 to 90½ per cent. Motors of this type operate at constant speed. Variable-speed a. c. motors will be considered later.

Reversing. To reverse the direction of rotation of induction motors, interchange leads as follows: For a two-phase four-wire motor, interchange the two leads of one phase; for a two-phase three-wire motor, interchange the two outside leads; for a three-phase motor, direction of rotation can be reversed by interchanging any two of the motor leads.

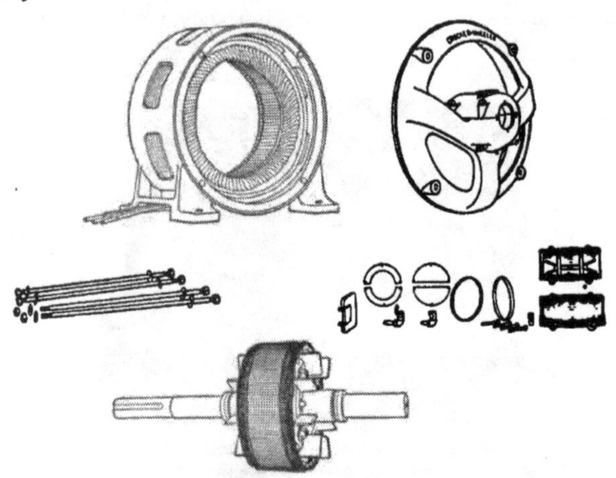

Fig. 47. Exploded View of Crocker-Wheeler Induction Motor

Output. The output of an induction motor varies with the square of the voltage at the motor terminals. If, for instance, the terminal voltage happens to be 15 per cent low (85 per cent of the rated voltage), a motor, which at the rated voltage gives a maximum of 200 per cent of its rated output, will give only $(85/100)^2 \times 200$, or 144 per cent of its rated output.

When an induction motor is overloaded, it draws an excessive current from the supply mains. The torque increases up to a certain value of slip. The rotor at this time has dropped back in position to such a point that the machine is unstable and, if the load is increased beyond this value, the machine comes to rest. This is called the "pull out" point of the motor. The speed of an induction motor varies directly with the frequency of the circuit on which it is operating.

Starting Compensator. It has already been pointed out that an induction motor of the squirrel-cage type can be started by simply closing the stator, or primary switch, but this calls for a very large rush of current. In order to cut down the starting

Fig. 48. Type NR General Electric Starting Compensator with No-Voltage and Overload Release with Enclosing Cover

current, when the maximum starting effort is not necessary, a starting compensator is usually employed. This compensator is an autotransformer, reducing the potential at the terminals of the motor and consequently diminishing the current taken by it.

Fig. 48 illustrates a compensator manufactured by the General Electric Company, and Figs. 49, 50, and 51 show the connections employed for two- and three-phase motors with compensators. Fig. 48 shows a compensator with overload relays mounted above it. These relays are arranged to trip the oil switch, when the current is excessive, and have time limit dashpots so that the switch will not trip during the starting operation. Fuses are sometimes used instead of these overload relays.

This type of compensator consists of coils (three for three-phase, two for two-phase) wound upon laminated iron cores, a spring return double-throw oil switch assembled in a metallic case, and a no-voltage release. To start the motor the switch should be thrown into the starting position with a quick firm thrust and held there until the motor comes up to speed (which usually requires but a few seconds) and then, with one quick firm movement, it should be pulled over into the running position, where it is held by a lever engaging with the no-voltage release mechanism.

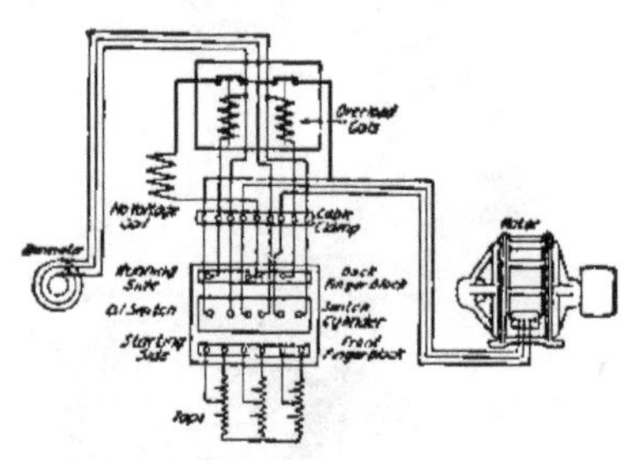

Fig. 49. Wiring Diagram for Three-Phase Starting-Compensator with Overload Release
Courtesy of General Electric Company

Never, in any case, should the motor be started by "touching," that is, by throwing the switch partly into the starting position and pulling it quickly out a number of times. This does not reduce the rush of current at starting, but, on the contrary, it produces a number of successive rushes in place of the one which it has attempted to avoid and, what is often a more serious matter, causes the contact fingers to be so badly burned that it is necessary to replace them.

When, however, the switch is properly thrown into the starting position, the current passes through the compensator. The windings of the compensator should be provided with a number of taps, so that the motor current may be obtained at a voltage considerably less than the line potential and thus start the motor with a minimum disturbance on the line.

With the switch in the running position, the compensator

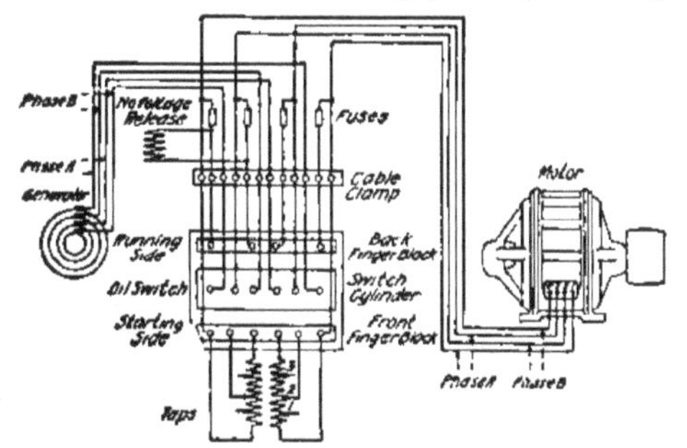

Fig. 50. Wiring Diagram for Two-Phase Starting Compensator for Four-Wire Circuit with Fuses

Courtesy of General Electric Company

windings are entirely disconnected and the motor takes its current directly from the lines at full potential.

Compensators are usually connected for low starting voltage

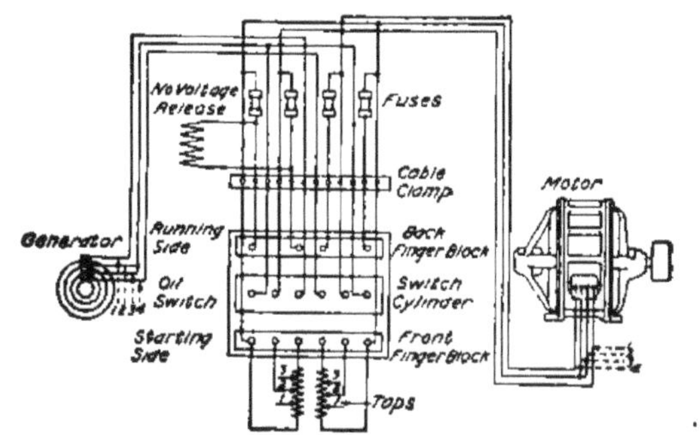

Fig. 51. Wiring Diagram for Two-Phase Starting Compensator for Three-Wire Circuit

Courtesy of General Electric Company

and torque. If the motor will not start its load with these connections, the next higher voltage taps should be tried, and so on until taps are found which will give the required torque. Motors of small size (say, five h.p. or under) are usually started by connecting directly to the line without resistance or compensator.

Polar-Wound Type. Another form of constant-speed induction motor is one in which the rotor is definite or polar wound.

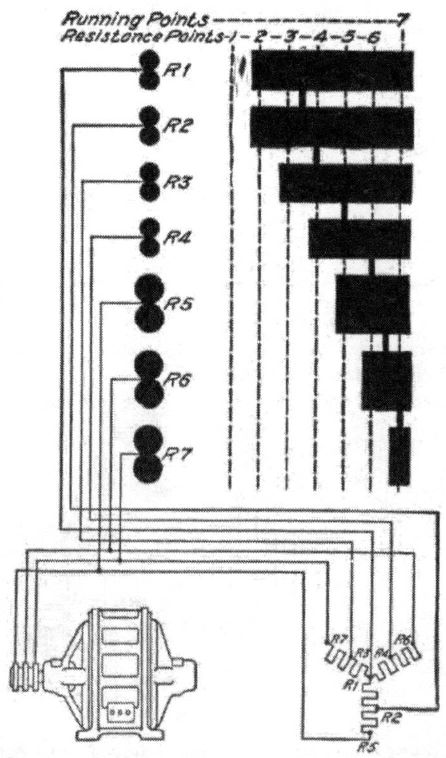

Fig. 52. Diagram of Simple Controller for Regulating the Resistance in the Rotor of a Slip-Ring Motor

The ends of this winding are connected to a resistance inside the rotor, arranged to be cut out while the machine is running, by a sliding contact, operated by a device on the motor shaft.

Starting. This type of motor is started without the use of a compensator in the leads to the stator to reduce the applied voltage. The device for cutting out the resistance is set so that all the resistance is connected in the rotor circuit when the motor is at rest. The stator, or primary circuit, is connected directly to the line and, as the motor attains speed, the successive steps of resistance are cut out until, on the last step, the resistance is short-circuited. The resistance must not be left in circuit after the motor has attained normal speed, since it is not intended for regulating the speed

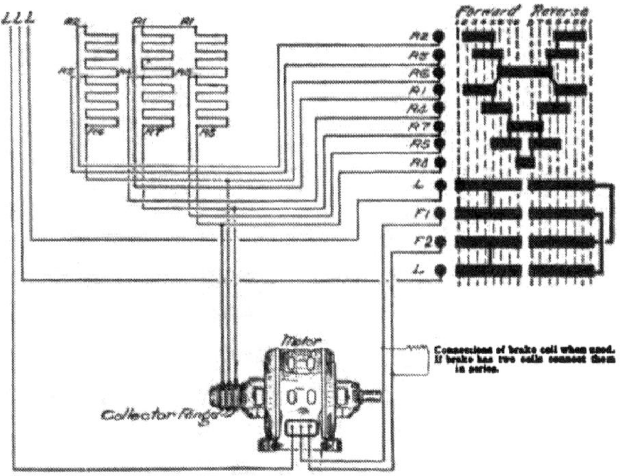

Fig. 53. Control Diagram for Slip-Ring Induction Motor

of the motor. This motor comes up to speed in about 15 to 18 seconds.

Motors of this type exert a high starting torque. For instance, with a starting current 1½ times full-load current, 1½ times normal full-load torque is obtained. The power factor and efficiency of these motors are about the same as for the squirrel-cage type.

Slip-Ring Type. Still a third type of polyphase motor is one which has the polar-wound rotor similar to the one previously described. However, the ends of the Y-connected rotor windings are brought out to three collector rings mounted on and insulated from the shaft. Brushes bear on these rings and connect to some form of Y-connected resistance.

Control. In most sizes of motor a controller, similar to the ordinary street-car controller, cuts out the resistance in the rotor circuit step by step, the resistance being entirely short-circuited

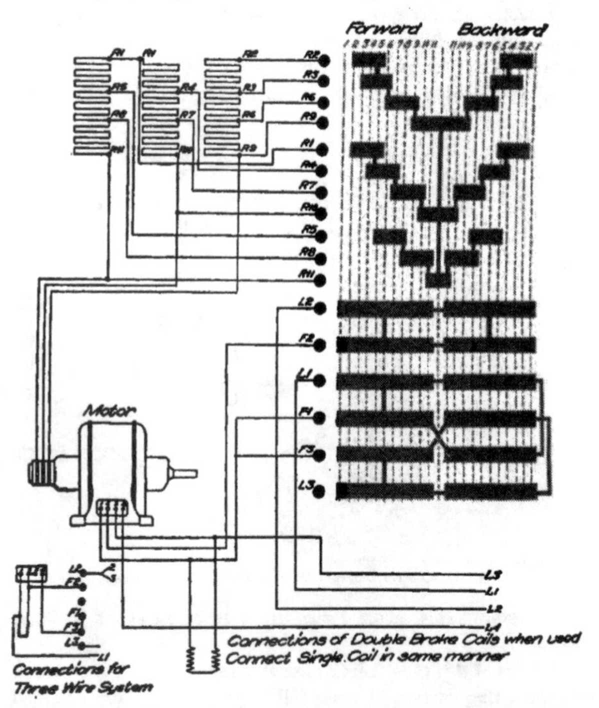

Fig. 54. Wiring Diagram for Control Mechanism of Slip-Ring Induction Motor

on the final step. The current to the stator or primary of the motor may also pass through the controller. Especially is this the case if the motor is to be operated in either direction. Figs. 52, 53, and 54 each show a method of control for the slip-ring type of induction motor.

In larger sizes of motor, the control is accomplished by a system of electrically-operated contactors which give the same sequence of connections as the manual control already shown. This type of motor may be operated at constant speed after all resistance is cut out of circuit, or it may be used as a variable-speed motor, in which case the resistance must be proportioned for continuous duty on any point of the control. Examples of this kind are elevators, cranes, hoists, and other applications requiring variable speed.

Fig. 55. Type KS Single-Phase Induction Motor
Courtesy of General Electric Company

Starting torque and current in this type of motor depend upon the resistance in the secondary circuit, full-load current giving approximately full-load torque.

Single-Phase Motors

Split-Phase Type. Owing to its simplicity, strength, and good electrical characteristics, the squirrel-cage polyphase motor is generally recognized as the ideal type for use on alternating-current power circuits. There are, however, many localities where for various reasons it becomes necessary for the central station to supply single-phase current to power consumers. The single-phase motor in sizes from about 1 to 15 h.p. takes care of such conditions.

Fig. 55 shows a General Electric single-phase motor, while Fig. 56 shows a Century Electric single-phase motor. This type of motor is suitable for constant-speed service where torque at starting does not exceed 140 per cent of full-load value.

Fig. 56. Exploded View of Century Electric Split-Phase Induction Motor

Starting. In starting, the rotor revolves freely on the shaft until approximately 75 per cent of the normal speed is

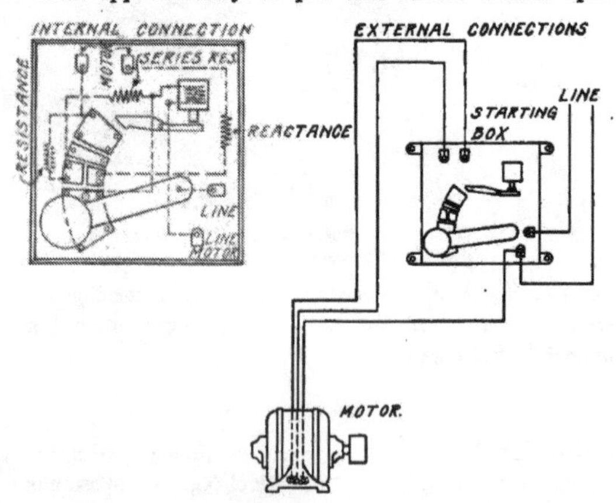

Fig. 57. Connections of Single-Phase Induction Motor with No-Voltage Release Starting Box

reached; the load is then picked up by the automatic action of a centrifugal clutch. The motor is started by means of a starting box containing resistance and reactance.

An induction motor designed to run on polyphase circuits will not start when its stator is connected directly to a single-phase circuit but, by providing the stator with a polyphase winding and "splitting the phase" of the applied single-phase current, sufficient starting torque is obtained. The current passing through the starting device will lag in the circuit having the reactance and cause the magnetic field to revolve. As soon as the motor has come up to speed, the connections to the stator are switched so that it is connected directly to the single-phase circuit. The motor will then continue to operate as a single-phase motor. Fig. 57 shows the connections necessary to operate the motor shown in Fig. 55. If a single-phase motor is started up by any external means and then thrown on the single-phase mains, it will continue to operate as a single-phase motor.

The efficiency of this type of motor at normal load varies from 70 per cent to 84 per cent, and the power factor from 67 per cent to 87 per cent in sizes of motors from 1 to 15 h.p.

Reversing. To reverse the direction of rotation, interchange any two of the three leads to the motor terminal block.

Repulsion Induction Motor. Another type of single-phase motor is that known as the "repulsion induction motor". The leading characteristics of the direct-current series-wound motor are well known. Operating through a wide range of speed and torque, this type has, however, no inherent speed regulation and its use is consequently confined either to fixed loads, like fans or pressure blowers, or to varying loads where the motor-controlling device is constantly under the operator's guidance. The speed, torque, and load characteristics of the series-commutator-type alternating-current motor being distinctly analogous to that of its direct-current prototype, the design fails to meet the requirements of constant-speed power service, this service demanding a motor which maintains good regulation after having once been brought up to speed, with torque values increasing as speed decreases; in other words, characteristics approaching those of the direct-current compound motor having the usual proportion of series-field winding.

The repulsion induction motor, however, gives this combination of series and shunt characteristics; that is, a limited speed and an increased torque with decrease in speed. In the straight repulsion motor, to secure the necessary starting torque, a direct-current armature is placed in a magnetic field excited by an alternating current and short-circuited through brushes set with a pre-

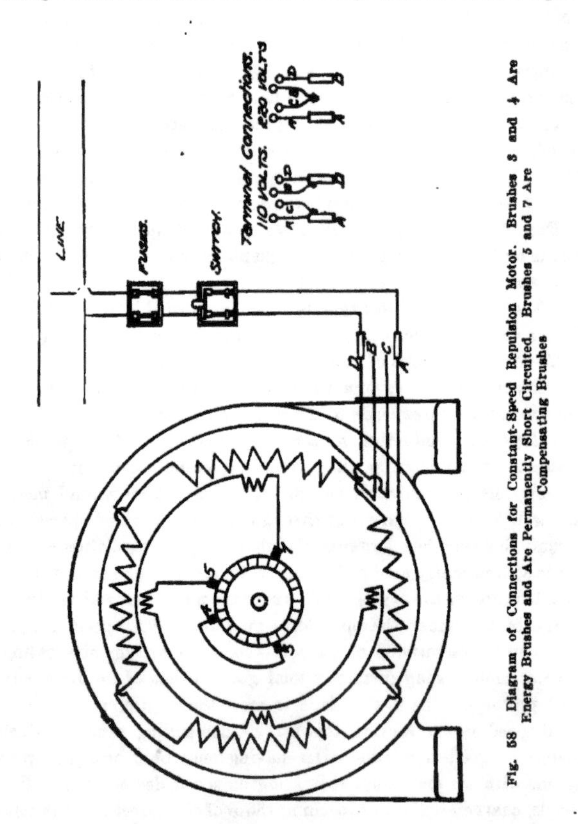

Fig. 58 Diagram of Connections for Constant-Speed Repulsion Motor. Brushes 3 and 4 Are Energy Brushes and Are Permanently Short Circuited. Brushes 5 and 7 Are Compensating Brushes

determined angular relation to the stator. To further improve the operating characteristics of the plain repulsion motor, a second set of brushes (*i. e.*, the compensating brushes) is placed at 90 electrical degrees from the main short-circuiting brushes (*i. e.*, the energy brushes), Fig. 58. The compensating field is auxiliary to

Fig. 59. Type RI Single-Phase Repulsion Induction Motor
Courtesy of General Electric Company

Fig. 60. Westinghouse Type AR Single-Phase Repulsion Motor

the main field and impresses upon the armature an electromotive force in angular and time phase with the electromotive force generated by the main field. In addition to correcting phase relation between the current and the voltage, thus giving approximately unity power factor at full load and power factors closely approaching unity over a wide range of load, the compensating field serves to restrict the maximum no-load speed and also permits, where varying speed service is involved, slight increase over synchronous values. The compensated repulsion motor is practically an induction motor capable of operation either above or below synchronous

Fig. 61. Induction Motor Driving Two Beaters Through Silent Chain
Courtesy of General Electric Company

speed, possessing heavy starting torque and high power factor at all loads as well as excellent efficiency constants. The motor has no tendency to spark or flash over, since the armature coils, successively short-circuited by the energy brushes, are not inductively placed in the magnetic field and have consequently only to commutate a low generated voltage. Figs. 59 and 60 show standard types of repulsion induction motors.

Starting Torque and Current. Starting torque is 200 or 250 per cent of rated-load torque with rated-line voltage. A starting current of 225 per cent of rated-load current is required to produce

ELEVEN 4000-K.W. 2200-VOLT GENERATORS. POWER HOUSE NO. 2. NIAGARA FALLS POWER COMPANY

rated-load torque. Where it is desired to reduce the starting current to the minimum, starting rheostats may be used.

The efficiency of the repulsion induction motor varies from 67 to 85 per cent and the power factor from 94 to 99 per cent, depending on the size of the motor.

The motor may be constant speed, constant speed reversible, variable speed, or variable speed reversible.

Typical Induction-Motor Installations. Induction motors of the three kinds described are extensively used in all manner of

Fig. 62. Induction Motor Belted to Berlin Sander
Courtesy of Westinghouse Electric and Manufacturing Company

industrial plants, the type of motor depending upon its applicability to the work at hand. Figs. 61 and 62 show induction motors in different industrial applications.

Synchronous Motors

As previously stated, the synchronous motor resembles the revolving-field generator, except that the field of the synchronous motor may be provided with a squirrel-cage winding. Fig. 63 shows a Westinghouse synchronous motor connected as a synchronous condenser.

Starting. This winding enables the synchronous motor to start up as an induction motor. Some device, such as a compensator, is used to reduce the line voltage, applied to the stator or primary of the motor, so as to reduce the rush of current in starting. Since the synchronous motor starts as a low-efficiency induction motor, it follows that it will not start under load, but will develop only enough torque to pull itself into synchronism.

The usual method of starting a synchronous motor is as follows: A starting compensator with several taps is provided, which

Fig. 63. 600 K.V.A. Self-Starting Synchronous Condenser
Courtesy of Westinghouse Electric and Manufacturing Company

may reduce the voltage applied to the stator of the motor to a value of from 50 to 70 per cent of normal line voltage, depending upon the amount of current required to bring the motor to about 95 per cent of synchronous speed. At this speed the field switch may be closed and the current in the field adjusted to about correct value. The full-line voltage may then be applied to the motor, when it will pull into synchronism. It will then be necessary to adjust the strength of the field to a value depending on whether the machine is to operate as a motor or as a synchronous condenser.

Some manufacturers recommend starting synchronous motors with the field short-circuited through a resistance. This is usually accomplished by making the discharge resistance heavy enough so that it will stand the induced current in the fields. In that case the field switch is left in the discharge clips during starting. The synchronous motor may be brought up to speed by an auxiliary starting motor and synchronized on the system, just as already described under the subject of a. c. generators.

Fig. 64. 1200 Kw. Synchronous Motor-Generator Set
Courtesy of Crocker-Wheeler Company

Speed. The speed of a synchronous motor depends only on the frequency and is constant for a given frequency up to a certain point. If loaded beyond this point, the motor "pulls out", or comes to rest.

Uses of Synchronous Motors. *Driving Generators.* The most usual application of synchronous motors is for driving d. c. or a. c. generators. They are largely used for driving d. c. generators for railway, lighting, and power service because of their good speed characteristics. In such cases they are usually direct-connected to the generator and mounted on the same base with it, Fig. 64. In cases where it is necessary to obtain current at a frequency differing from the supply circuit, synchronous motors of one frequency are connected to a. c. generators of another frequency. Such combinations are called "frequency changer sets". Where the machines are direct-connected and therefore have the same speed, the ratio of poles on the two machines must correspond to the ratio of frequencies desired.

Fig. 65. General Electric Synchronous Condenser with Direct-Connected Exciter

Driving Compressors, Blowers, Etc. Synchronous motors are also well adapted for direct connection to some types of machines, such as air compressors and blowers. The speeds which can be obtained are well suited to such machines and considerable fly-wheel effect can be obtained, which is also of advantage.

Synchronous Condensers. Another very useful application of synchronous motors is their use as synchronous condensers. In any of the applications mentioned above, the field strength of the motor can be increased until the motor is taking leading current from the line. Thus instead of decreasing the general power factor of the system, as is the case with the induction motor, these motors may be made to increase the power factor of the system.

Where the motor is driving other load, this corrective effect must necessarily be but a small part of the total capacity of the machine. If the motor be floated on the line without carrying any mechanical load except its own rotor, a proportionally larger corrective effect can be obtained. A machine operated in this way is called a "synchronous condenser". In many installations having a large number of induction motors with a low average load, the power factor is very low. The increased capacity made available from the generators by increasing the power factor will often warrant a considerable investment in synchronous condensers in such cases. Fig. 65 shows a General Electric synchronous condenser with exciter which may be used under such conditions. The sole purpose of these machines is to raise the general power factor of the system.

OPERATION

Directions for Running Generators and Motors. *Preliminary Run with No Load.* If possible, a new machine should be run with no load or with a light one for several hours. It is bad practice to start a new machine with its full load or even a large fraction of it. This is true even if the machine has been fully tested by its manufacturer and is apparently in perfect condition, because there may be some fault produced in setting it up, or some other circumstance that would cause trouble. All machinery requires some adjustment and care for a certain time to get it into smooth working order.

Voltage and Current Regulation. A generator requires that its voltage or current should be observed and regulated if it varies. The attendant should always be ready and sure to detect and overcome any trouble, such as sparking, heating, noise, abnormally high or low speed, etc., before any injury is caused. Such directions should be thoroughly committed to memory in order promptly to detect and remedy any trouble when it occurs suddenly, as is usually the case. If possible, the machine should be shut down instantly when any indication of trouble appears, in order to avoid injury and to give time for examination.

Keep Tools Away from Machines. Keep all tools or pieces

of iron or steel away from the machine while running. Otherwise, they might be drawn in by the magnetism, perhaps getting between the armature and pole pieces, thus ruining the machine.

Commutator and Brushes. Particular care should be given to the commutator and brushes, so that the former is kept perfectly smooth and the latter are in proper adjustment. Avoid lifting brushes when machine is operating, unless there are several in parallel.

Bearings. Touch the bearings occasionally to see whether or not they are hot. To determine whether the armature is running hot, place the hand in the current of air thrown out from it by centrifugal force.

Overloading. Special care should be observed by anyone who runs a generator or motor, to *avoid overloading* it, because this is the cause of most of the troubles which occur.

Personal Safety. The matter of personal safety is of great importance in the installation, care, and management of dynamo-electric machinery, both from the humanitarian and from the financial standpoint.

Precautions in Handling the Circuit. The safest rule is never to touch any conductor carrying current, and never to allow the body to form part of an electric circuit, no matter what the voltage. This, of course, is a rule which cannot be followed strictly in practice. However, every precaution should be taken to prevent accidents, and every device which adds to the personal safety of the men should be employed. Rubber gloves, rubber shoes, or both, should be used in handling circuits of 500 volts or over. Also these articles should frequently be tested. Tools with insulated handles, or a dry stick of wood, should be used instead of the hand for handling the wires. It should always be remembered that a wire may be "alive" through some unknown change in connection or through accidental contact with another wire, even when it is thought to be "dead".

High Voltages. On the high a.c. voltages now so common, even the above precautions are not sufficient. No work can ever be done on such circuits unless they are entirely disconnected from all sources of power. In addition, the wires should be thoroughly grounded before being touched. In grounding, the ground connection should be first made and last disconnected.

Stopping Generators or Motors. *Operating Alone.* A generator operating alone on a circuit can be slowed down and stopped without touching the switches, brushes, etc., in which case the current gradually decreases to zero; and then the connections can be opened without sparking or any other difficulty.

Operating in Parallel. However, when a generator is operating in parallel with other sources of power, it must not be stopped or reduced in speed until it is entirely disconnected from the system. Furthermore, the current generated by it should be reduced nearly to zero before its switch is opened. For a d.c. machine this is accomplished by adjusting the field rheostat of the machine to be cut out, great care being taken that the change is gradual. If the reduction be rapid, the voltage of the machine may drop so as to cause a back current. For a.c. generators the load is reduced by adjusting the engine governor to reduce the input. Field rheostats should not be changed on a.c. generators.

Constant-Current Generator. A constant-current generator may be cut into or out of circuit in series with others, and can be slowed down or stopped; or its armature or field coils may be short-circuited to prevent the action of the machine, without disconnecting it from the circuit. *It is absolutely necessary, however, to preserve the continuity of the circuit,* and not to attempt to open it at any point, as this would produce a dangerous arc. Hence, a by-path must be provided by closing the main circuit around the generator, before disconnecting it. This same rule applies to any lamp, motor, or other device on a constant-current system.

Never, except in an emergency, should any circuit be opened when heavily loaded; the flash at the contact points, the discharge of magnetism, and the mechanical shock are all decidedly objectionable.

Constant-Potential Motor. A constant-potential motor is stopped by turning the starting-box handle back to the first position; or, *if there is a switch, or circuit breaker, in the circuit,* as there always should be, it should be opened, after which the starting-box handle is moved back to be ready for starting again.

Immediately after a machine is stopped it should be thoroughly cleaned and put in condition for the next run. When not in use, machines should be covered with some waterproof material.

INTERIOR OF GENERATING STATION OF THE ONTARIO POWER COMPANY, NIAGARA FALLS, ONT.
The generators are direct-connected to horizontal double turbines.

MANAGEMENT OF DYNAMO-ELECTRIC MACHINERY

PART II

INSPECTION AND TESTING

MECHANICAL TESTS

Adjustment. Adjustment and the other points which depend merely upon mechanical construction are hardly capable of being investigated by a regular quantitative test, but they can and should be determined by a thorough inspection. In fact, a very careful examination of all parts of a machine should always precede any test. This should be done for two reasons: *First,* to get the machine into proper condition for a fair test, and, *second,* to determine whether the materials and the workmanship are of the best quality and satisfactory in every respect. A loose screw or connection might interfere with a good test; and a poorly fitting bearing, brush holder, or other part might show that the machine was badly made.

If it is necessary to take the machine apart for cleaning or inspection, the greatest care should be exercised in marking, numbering, and placing the parts, in order to be sure to get them together exactly the same as before. In taking a machine apart or putting it together, only the minimum force should be used. Much force usually means that something wrong is being done. A wood or rawhide mallet is preferable to an iron hammer, since it does not bruise or mar the parts. Usually screws, nuts, and other parts should be set up fairly tight, but not tight enough to run any risk of breaking or straining anything. Shaking or trying each screw or other part with a wrench or screw driver, will show whether any of them are too loose or otherwise out of adjustment.

Friction. *Revolving Rotor by Hand.* The friction of the bearings and brushes can be tested roughly by revolving a small armature or rotor by hand, or slowly by power for a large one, and noting if it requires more than the normal amount of force. Excessive friction is easily distinguished, even by inexperienced persons. Another method is to revolve the armature and see if it continues to revolve by itself freely for some time. A well-made machine in good condition and running at or near full speed will continue to run for several minutes after the turning force is removed.

Revolving by Motor. Another method for measuring the friction of a machine is to run it by another machine used as a motor, and determine the volts and amperes required, *first,* with brushes lifted, and, *second,* with brushes on the commutator with the usual pressure. The torque or force exerted by the driving machine is afterwards measured by a Prony brake in the manner described hereafter for testing torque, care being taken to make the Prony brake measurements at exactly the same volts and amperes as were required in the friction tests. In this way the torques exerted by the driving machine to overcome friction in each of the first two tests are determined, and these torques, compared with the total torque of the machine being tested, should give percentages not exceeding about 2 per cent with the brushes up, or 3 per cent with the brushes down. The magnetic pull of the field of the armature may be very great, if the latter is not exactly in the center of the space between the pole pieces. This would have the effect of increasing the friction of the shaft in the bearings when the field is magnetized. It occurs to a certain extent in all cases but it should be corrected, if it becomes excessive. This may be tested by turning the current into the fields, being sure to leave the armature disconnected, and then determining the friction. The friction in this case should not be more than 2 to 4 per cent.

Tests for friction alone should be made at low speed, because at high speeds the effects of eddy currents and hysteresis enter and materially increase the apparent friction.

Balance. Perfection of balance of the revolving member of a machine is essential. On machines of small or medium size the revolving element should be laid on carefully leveled knife-edge steel ways. The heaviest side of the rotor will naturally assume the lowest position. Balance weights should then be placed diametrically opposite this point until the proper balance is obtained. It does not follow that the static balance will always be the correct running balance, and additional balancing may have to be done with the machine running at normal speed.

On very large machines, with which the preceding method can not conveniently be used, trial balance weights will have to be used until, by repeated tests, the proper position for the correct balance weights is obtained.

Noise. This cannot well be tested quantitatively, although it is very desirable that a machine should make as little noise as possible. Noise is produced by various causes. The machine should be run at full speed, and any noise and its cause carefully noted. A machine—especially the commutator—will nearly always run more quietly after it has been in use a week or more and has worn smooth.

Heating. *Resistance Method.* The most accurate way to determine the temperature rise in electrical apparatus is by measurements of resistance, before and after operating, for a specified time (usually 4 to 8 hours) under rated load. The rise of temperature is the percentage increase of initial resistance by the temperature coefficient for the initial temperature expressed in per cent. The standard room temperature or ambient temperature of reference was formerly 25 degrees centigrade but it is now fixed at 40 degrees centigrade, corresponding to high temperatures in engine rooms and in hot weather. For ordinary tests it may be assumed that the resistance of copper increases .4 per cent for each degree centigrade rise in temperature, this being correct for 15 degrees centigrade, being .385 per cent at 25 degrees centigrade and .364 per cent at 40 degrees centigrade.

Thermometer Method. Thermometers are also largely used for determining the temperature rise in apparatus. The indications obtained by thermometer may be about 5 degrees centigrade lower than those obtained by the resistance method. When resistance is taken by thermometer, great care should be taken to prevent radiation. The thermometer should be covered with a pad of waste or cloth, not exceeding two inches in diameter, and allowance should always be made for the fact that the temperature is always less at the surface than at the interior.

The American Institute of Electrical Engineers has made

general recommendations as to the limits of temperature advisable with various kinds of insulating material. Other authorities state the limits allowable for various types of machinery, for example, open and enclosed.

ELECTRICAL TESTS

WIRING

Before any work is done in connection with the installation of wires or cables, complete plans should be drawn, showing the proper relative location of all apparatus, the size of the conductors in every case, and the lengths of the runs. The path of the wires will depend on the system of wiring adopted but, in any case, it should be as simple and straightforward as possible. The rules of the National Board of Fire Underwriters should be closely adhered to, as well as the rules of any local board having jurisdiction. It is a good plan to have any doubtful points passed on by an inspector before the work is completed if the expense involved is very great. Otherwise, the work may have to be done over after completion, which more than doubles the expense. In general, the Underwriters recognize two methods of wiring—"exposed" and "concealed".

Exposed Wiring. Exposed wiring on cleats and knobs or, for large cables, on cable racks with porcelain insulators is used less today than ever before. It has the great advantage of cheapness and accessibility, which will always make it popular in some places. However, the wires have an unsightly appearance and are subject to injury from sources external to themselves which greatly multiply the chances for trouble.

Concealed Wiring. The use of concrete construction in power plants of all kinds, and in industrial plants as well, has made it possible to make practically all wiring in such buildings concealed in iron, clay, or fiber conduits. Placing the conduits in the floors or walls puts the wires entirely out of sight, where they are not subject to injury from external sources, and where they are usually convenient for connection to the apparatus. Especially for power houses this method has become practically universal.

Houses, Office Buildings, Etc. The wiring of houses, office buildings, factories, etc., is practically all placed in conduits, as stated above. The voltage in all such cases is low; that is, usually not over 600 volts. The wires are insulated with rubber, paper, or varnished cambric of the thickness necessary for the voltage used, and are covered with braid if installed in dry locations. If moisture is likely to be encountered, the wires are covered with a lead sheath.

Power Houses. In power houses similar construction is used for voltages up to and including 13,200 volts. For busbars, especially when enclosed in fireproof compartments, bare bar or rod is usually used. For voltages above 13,200 there are many installations where the wires are insulated even up to 33,000 volts. However, insulation for such potentials is considered by most engineers as false protection. That is, it is not safe to depend upon the insulation to protect one against shock, and its presence is apt to give to the operator a false sense of security which might cost him his life. It has, therefore, become an almost universal practice to make all station wiring for high voltages of bare wire mounted on porcelain insulators. With such construction *it is evident* that one must never come near the wires, and the result is that the wiring is safer than if it were insulated.

When alternating-current conductors are enclosed in iron conduits, both wires of each phase, or all the wires, must be run in the same duct; otherwise the inductance would be excessive.

Size of Conductors. All conductors, including those connecting the machine with the switchboard, as well as the busbars on the latter, should be of ample size to be free from overheating and excessive loss of voltage. The voltage drop between the generator and the switchboard should not exceed one-half per cent at full load. Excessive drop at this point interferes with proper regulation and adds to the less easily avoided drop on the distribution system.

The safe-carrying capacities of copper conductors as recommended by the National Board of Fire Underwriters are given in Table II.

TABLE II

Safe-Carrying Capacities of Copper Wires

Rubber Insulation		Other Insulations	
B. & S. G.	Amperes	Amperes	Circular Mils
18	3	5	1,624
16	6	8	2,583
14	12	16	4,107
12	17	23	6,530
10	24	32	10,380
8	33	46	16,510
6	46	65	26,250
5	54	77	33,100
4	65	92	41,740
3	76	110	52,630
2	90	131	66,370
1	107	156	83,690
0	127	185	105,500
00	150	220	133,100
000	177	262	167,800
0000	210	312	211,600
Circular Mils			
200,000	200	300	
300,000	270	400	
400,000	330	500	
500,000	390	590	
600,000	450	680	
700,000	500	760	
800,000	550	840	
900,000	600	920	
1,000,000	650	1,000	
1,100,000	690	1,080	
1,200,000	730	1,150	
1,300,000	770	1,220	
1,400,000	810	1,290	
1,500,000	850	1,360	
1,600,000	890	1,430	
1,700,000	930	1,490	
1,800,000	970	1,550	
1,900,000	1,010	1,610	
2,000,000	1,050	1,670	

The lower limit is specified for rubber-covered wires to prevent gradual deterioration of the high insulations by the heat of the wires, but not from fear of igniting the insulation. The question of drop is not taken into consideration in the above tables.

The safe-carrying capacity of insulated aluminum wire is 84 per cent of that given for copper wires of corresponding size and insulation.

PROTECTIVE APPARATUS

Switches. Switches are devices for closing or opening a circuit. Several different kinds are used, the design depending upon the character and the severity of the service. For low voltage circuits (125 or 250 volts) of 10 amperes or less, snap switches are often used. These switches are usually of the kind mounted on porcelain bases and enclosed with a metal cover. The handle operates on a spring which in turn operates the contact blade or blades. Thus the contacts remain in position until the spring pressure overcomes the friction, when the switch opens or closes with a quick snapping action.

Fig. 66. Exploded View of Double-Pole Lever Switch
Courtesy of General Electric Company

Lever Switches. Lever or knife switches consist of blades of copper operating in clips attached to copper blocks. One end of the blade is hinged in the clips and spring washers are provided to insure pressure enough for good contact and to take up any wear between blade and clips. An insulating handle is attached to the other end of the blade. If the switch has two or more poles, they are fastened together by means of an insulating crossbar, and the handle is attached to the crossbar instead of to the blade direct. Fig. 66 shows such a lever switch for mounting on a switchboard. It also shows an exploded view of the parts.

The general construction, spacing, etc., is, in this country, fixed by the rules of the National Board of Fire Underwriters.

Fig. 67. Complete Panel Board in Steel Cabinet

Their rules require that switches have sufficient metal for the neces sary mechanical strength, and that the temperature rise shall not be more than 28 degrees centigrade above the surrounding air, when carrying full rated load continuously. With good contacts they allow a rated current density of 75 amperes for each square inch of contact area. Switches of over 30 amperes must be provided with terminal lugs screwed or bolted into the switch into which the conducting wires must be soldered. In practice, lever switches of less than 30-ampere capacity are seldom manufactured and, on that size, lugs are usually furnished.

Lever switches should be mounted on non-absorptive non-combustible insulating bases or panels such as slate, marble, or porcelain. Where switches are mounted on bases which are to be placed against a flat surface, the connection must be made at the front of the switch. In this case the block is extended far enough so that the terminal can be screwed or bolted on to the face of the block. Where the switches are mounted on switchboard panels, a stud is passed through the panel from the switch block and the terminals are fastened to it by means of nuts.

Lever switches should be mounted so that gravity tends to open them rather than close them. They should be grouped as far as possible for ease of operation. They should be easily accessible and located in a dry place. Modern practice requires that, for small capacities and especially when in exposed places, the group of switches be mounted on a panel board and placed in a steel cabinet properly lined with some material such as slate, Fig. 67.

Plain lever switches are not used for opening circuits under load on more than 300 volts d. c. or 500 volts a. c. However, they may be used on circuits of all voltages for disconnecting purposes at no load or very small load and where any considerable current is ruptured by some other device.

The Underwriters' Rules cover a large number of points of design and construction not mentioned above, but they are of interest principally to the manufacturers, and all reputable manufacturers follow them.

Oil Switches. For alternating-current circuits, especially where the voltage is high, lever switches are not suitable for operation under load. An oil switch is one in which the circuit is made or broken under oil. Fig. 68 shows a single-pole 15,000-volt oil switch for switchboard service, with the oil tank removed. Such switches are mounted back of, or remote from, the panels, and are operated by a handle connected to the switch through a system of links and levers. These switches usually open by gravity, either when tripped by hand or automatically. When tripped automatically by abnormal conditions in the circuit, they are sometimes called "oil circuit breakers". They are manufactured with one or more poles, sometimes with one tank for all poles and sometimes with a separate tank for each pole.

Switches built along the same general lines are also manufactured for mounting on walls or on motor-driven machines and are used for controlling individual motors. In textile mills and other places where the fire risk is great, such switches are almost essential even on low-voltage circuits.

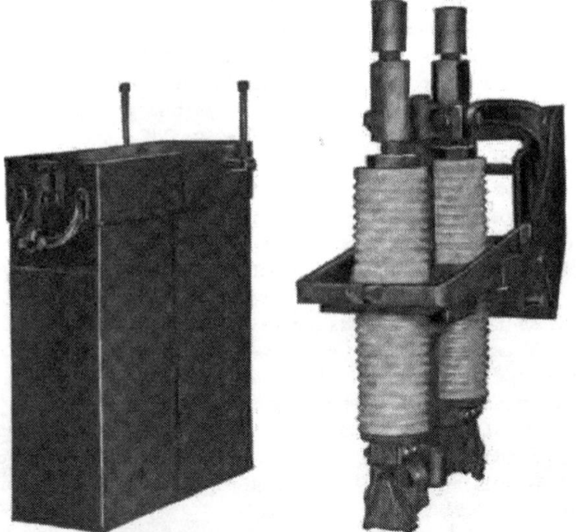

Fig. 68. General Type of Oil Switch with Cover Removed
Courtesy of General Electric Company

Oil switches are not suitable for use on direct-current circuits, since there is no reversal of current and the arc is much harder to extinguish than the a. c. arc.

Circuit Breakers. Where large currents are to be broken, when the voltage is high, or where automatic features are required, the usual direct-current practice is to use circuit breakers. For electric railway service on cars, or in places where the equipment must be totally enclosed, magnetic blowout circuit breakers are still used. These breakers are arranged so that, as the circuit is broken, current is caused to flow in a coil which produces a magnetic field so placed as to draw out the arc and cause it to break quickly.

The most common form of circuit breaker is the carbon-break circuit breaker as illustrated in Fig. 69. This breaker has a copper-

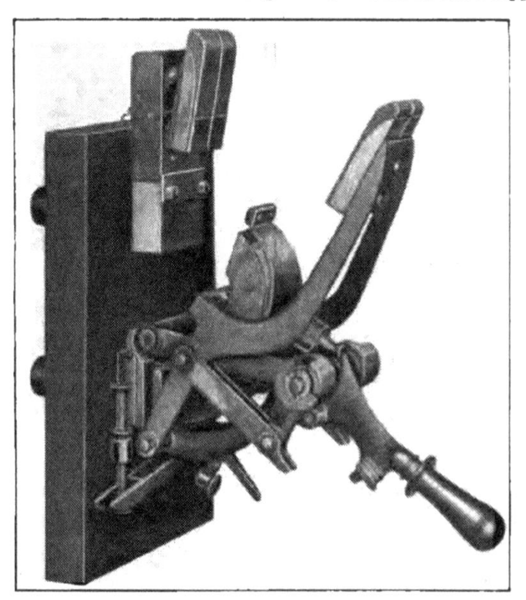

Fig. 69. "Type C" 650-Volt 2000-Ampere Single-Pole Overload Circuit Breaker
Courtesy of General Electric Company

leaf brush for continuously carrying the current. As the circuit breaker is opened, or opens automatically, this brush leaves contact first. The current is then temporarily carried by a secondary copper contact which opens second. Last of all the carbon contacts open and actually break the circuit. This arrangement insures that the burning will all come on the renewable carbon block and the main contacts will always be left in good condition.

Automatic overload tripping is obtained by arranging a magnetic circuit in such a way that at a predetermined current value an armature will be lifted and, in its movement, will strike a tripping latch which in turn releases the moving element of the breaker so that it opens by gravity and the spring action of the contacts. Other automatic features can also be obtained by means of auxiliary attachments or separate relays acting on the circuit-breaker mechanism in a similar way. Thus the circuit breaker may be tripped by an underload, by low voltage, or by the reversal of current, when desired.

Carbon-break circuit breakers are often used in alternating-current circuits for voltages of 600 and below. They are of the same general appearance as those used in direct-current circuits, and operate on the same principle.

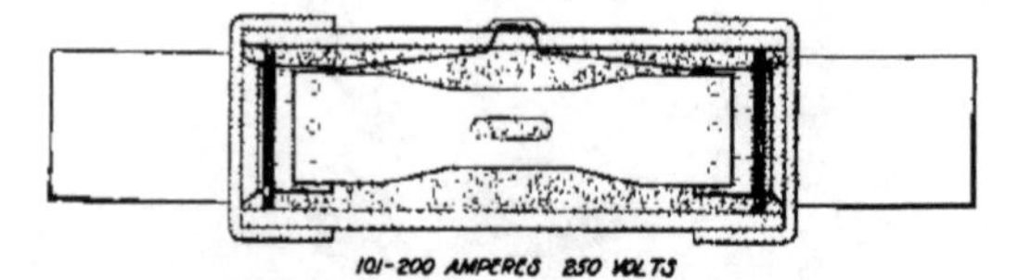

Fig. 70. Section of Enclosed Fuse

Fuses. Because of the great expense of replacements on circuits where trouble is frequent and severe, fuses are being abandoned in favor of circuit breakers for all large power circuits. However, there are so many places where fuses have the advantage, because they are simpler, that they may still be considered among the most important protective devices used.

A fuse is merely a strip of metal, inserted in a circuit, of such cross section that at a given current it will melt, thus opening the circuit. If this fuse is made so that its ultimate carrying capacity is less than the safe-carrying capacity of the rest of the circuit, it will protect the apparatus against the excessive current. This in general describes the principle and function of the fuse.

The form which a fuse may assume depends upon the severity of the service, the capacity of the circuit, and the voltage. Figs. 70, 71, and 72 show, respectively, a section of a National Electric Code

Standard enclosed fuse, an open-link fuse, and an expulsion fuse for higher voltage a. c. circuits.

Open-link fuses are, as the name implies, merely links of metal with limited current-carrying capacity, which are mounted between two terminal blocks and fastened to them by bolts or screws. The metal used is usually an alloy of special composition for the service. The National Board of Fire Underwriters requires that the fuses be stamped with about 80 per cent of the maximum current which they can carry indefinitely, thus allowing about 25 per cent overload before the fuse melts. The minimum current which will melt the fuse in about five minutes is taken as the melting point.

Fig. 71. Open-Link Fuse

It can readily be seen that on circuits of large capacity, or circuits having a large generating capacity back of them, the opening of an open-link fuse would produce a heavy flash and a dangerous splashing of metal. The use of enclosed fuses on circuits of from 30- to 600-ampere capacity has become almost universal and this practice is to be recommended in any case where fuses are obtainable which meet the rules of the Underwriters. The Underwriters' Rules cover the general construction, performance, and dimensions of fuses within the range of capacities given above.

Fig. 72. General Electric Single-Pole Expulsion Fuse Block

Fuses should always be employed when the size of the wire changes, or where connections between any electrical apparatus and the conductors are made. They must be mounted on slate, marble, or porcelain bases; and all metallic fittings employed in making electrical contacts must have sufficient cross section to insure mechanical stiffness and carrying capacity.

For higher voltage circuits used in alternating current, the open fuse has the same objections as for direct current and in a more marked degree. The completely enclosed fuse is also found unsatisfactory. It has, therefore, been necessary to adopt a different principle of operation for these fuses. The link of metal used for the fuse is enclosed in a tube which has an enlarged section at one end and is open at the other end. The reduced section of the fuse is placed so as to come within the enlarged section, or explosion chamber of the tube. When the fuse is ruptured, gases form in the explosion chamber which blow out through the tube, thus opening the circuit quickly and effectively.

As a general proposition, fuses of any kind are superfluous or undesirable where the conditions warrant the use of automatic circuit breakers, as in power houses, sub-stations, and factories. The conditions which should be met in the manufacture of a good fuse are as follows:

(1) They should melt at a definite current.

(2) The melting point should not change due to time, current, heating, or any reasonable service conditions.

(3) They should act quickly at the current for which they are marked.

(4) They should give firm and lasting contacts with the terminal blocks to which they are attached.

While the above conditions are hard to meet and perfect fuses cannot be expected, they do give a fair degree of protection and will continue to be used for many classes of service.

ELECTRICAL RESISTANCE

Among the most important tests which it is necessary to make in connection with dynamo-electric machinery are those for resistance. There are two principal classes of resistance tests that must be made in connection with generators and motors: *first*, the resistance of the wires or conductors themselves, called the *metallic* resistance; and *second*, the resistance of the insulation of the wires, known as the *insulation* resistance. The latter should always be as high as possible, because a low insulation resistance not only allows current to leak, but also causes "burn-outs" and other accidents. Metallic resistance, such, for example, as the resistance of the armature or field coils, is commonly tested either by the Wheatstone bridge or by the "drop" (fall-of-potential) method.

Metallic Resistance

Wheatstone Bridge Method. The Wheatstone bridge is simply a number of branch circuits connected as indicated in Fig. 73. *A, B,* and *C* are resistances the values of which are known. *X* is the resistance which is being measured. *G* is a galvanometer, *S* its key, and *E* is a battery of one or two cells controlled by a key *K,* all being connected as shown. The resistance *C* is varied until the galvanometer shows no deflection when the keys *K* and *S* are closed in the order named. If the key *S* should be closed before *K*, or at the same moment, the inductive effect might produce a pronounced deflection of the galvanometer needle, and thus probably cause confusion. The value of the resistance *X* is then found by multiplying together resistances *C* and *B,* and dividing by *A;* that is,

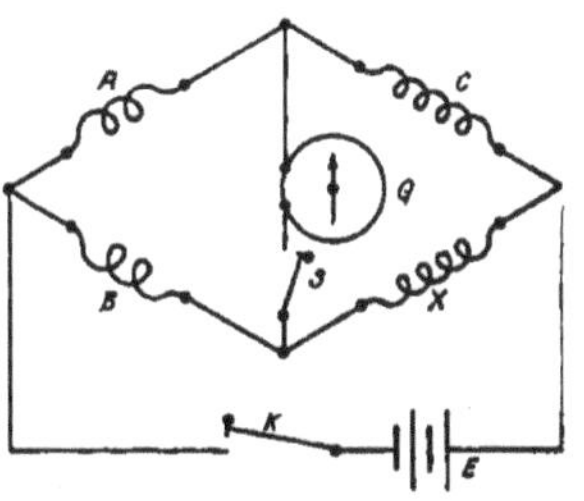

Fig. 73. Wheatstone Bridge Diagram

$$X = \frac{CB}{A}$$

Fig. 74. Portable Wheatstone Bridge
Courtesy of Roller-Smith Company

Portable Bridge. A very convenient form of this apparatus is what is known as the portable bridge, Fig. 74. This consists of a plug resistance box of the decade type, corresponding to *C* of Fig. 73; the ratio coils corresponding to *A* and *B* of Fig. 73, four of these coils being multipliers and four dividers, giving a theoretical range from .001 ohms to over 11 megohms; sensitive galvanometer, battery, keys, etc. The resistance coils are of zero temperature coefficient and are noninductively wound. For the method of operation, the reader is referred to the article on Electrical Measurements. The Wheatstone bridge may be used for testing the resistances of almost any field coils that are found in practice. In measuring field resistances with the bridge, care must be taken to wait a considerable time after pressing the battery key, before pressing the galvanometer key, in order to allow time for the self-induction of the magnets to disappear.

Where the resistance of a generator armature, say, is too high for accurate bridge measurements, recourse may be had to an ohmmeter, or a bridge megger, both instruments of which are described in the article on Electrical Measurements. The method described in the following paragraph is also available.

Drop, or Fall-of-Potential, Method. The fall-of-potential method is well adapted for locating faults quickly and for testing the armature resistance of most generators and motors, or the resistance of contact between commutator and brushes, or other resistances which are usually only a few hundredths or even thousandths of an ohm. This method consists in passing a current through the armature and connections and a known resistance of say 1/100 ohm, all connected in series, as represented in Fig. 75. The "drop" or fall-of-potential in the armature and that in the known resistance are compared by connecting a voltmeter first to the terminals of the known resistance (marked *1* and *2*), and then to various other points on the circuit, as indicated by the dotted voltmeter terminals at *M, N, O, Q, R,* and *S,* so as to include successively each part to be tested. The deflections in all cases are directly proportional to the resistances included between the points touched by the terminals. The current needed depends upon the resistance of the circuit and the sensitiveness of the voltmeter. A resistance is used for limiting the current, which may be obtained from an ordinary 110- or 220-volt source of d.c. supply.

Instead of using a known resistance, an ammeter may be inserted in series with the resistance to be tested, the latter being then determined by Ohm's law, viz., *If E is the voltmeter deflection and I represents the amperes flowing, the resistance of the part under test is*

$$R=\frac{E}{I}$$

A "station" or a portable voltmeter may be used for the readings, and its terminals may be held in the hands, or they may be conveniently arranged to project from an insulating handle like a two-pronged fork. Usually 10 to 100 amperes and a low-reading voltmeter are needed for low resistances.

It is well to start with a small testing current, increasing it until a good deflection is obtained on the voltmeter. If a current of several amperes cannot be had, a few cells of storage battery or some strong primary battery can be used with a galvanometer or low-reading voltmeter.

The diagram indicates the testing of a machine with series fields. Shunt fields, on account of their high resistance, may be tested by the bridge method, or by voltmeter and ammeter read-

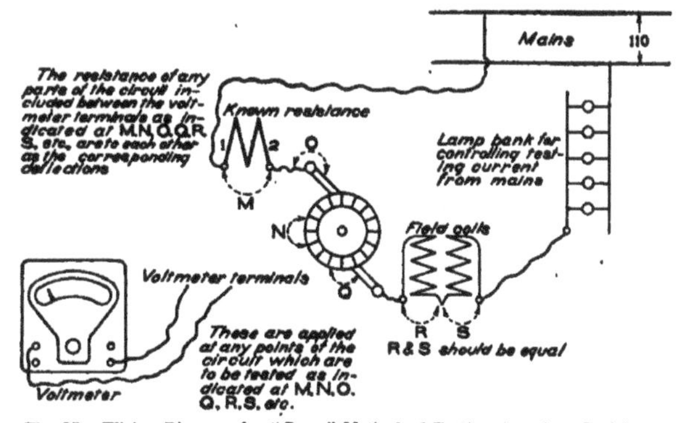

Fig. 75. Wiring Diagram for "Drop" Method of Testing Armature Resistance

ings; while the armature can be tested, as in Fig. 75. Of course it must not be allowed to revolve while its resistance is being tested.

The drop method of testing is also very useful in locating any fault. The two wires leading from the voltmeter are applied to any two points of the circuit, as indicated by the dotted lines, Fig. 75—for instance, to two adjacent commutator segments, or to a brush tip and the commutator; any break or poor contact will be indicated immediately by the deflection being larger than at some other similar part. This shows that the fault is between the two points to which the wires are applied. Thus, by moving these along on the circuit, the exact location of any irregularity, such as a bad contact, short circuit, or extra resistance, can be found.

Insulation Resistance

The insulation resistance of a generator or motor, that is, the resistance between its conductors and its frame, should be sufficiently high so that not more than one-millionth of its rated current will pass through it at normal voltage. Any value over one megohm in such cases is usually high enough. It is therefore beyond the range of ordinary Wheatstone bridge tests; but two good methods are applicable, the "direct-deflection" method and the voltmeter method.

Direct-Deflection Method. The direct-deflection method is carried out by connecting a sensitive galvanometer, such as a Thomson high-resistance reflecting galvanometer, in series with a known high resistance, usually a 100,000-ohm rheostat, a battery,

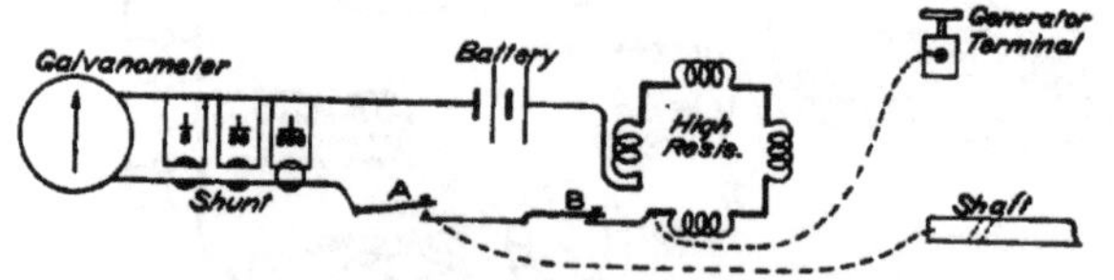

Fig. 76. Wiring Diagram for Direct-Deflection Method

and keys, as shown in Fig. 76. The galvanometer should be shunted with the 1/999 coil of the shunt, so that only 1/1000 of the current passes through the galvanometer, the machine being entirely disconnected. The keys A and B are closed and the steady deflection noted. It is well to use but one cell of the battery at first, and then increase the number, if necessary, until a considerable deflection is obtained. The circuit is then opened at the key B and connected by wires to the binding post or commutator and to the frame or shaft of the machine, as indicated by dotted lines, so that the machine insulation resistance is included directly in the circuit with the galvanometer and the battery. The key A is then closed and the deflection noted. Probably there will be little or no deflection on account of the high insulation resistance; and the shunt is changed to 1/99, 1/9, or left out entirely, if little deflection is obtained. In changing the shunt, the key should always be open; otherwise the full current is thrown on the galvanometer. The insulation is then calculated by the formula:

$$\text{Insulation resistance} = \frac{DRS}{d}$$

in which D is the first deflection without the machine connected; d is the deflection with the machine insulation in the circuit; R is the known high resistance; and S is the ratio of the shunted current through the galvanometer. If the shunt is 1/999 in the first test, and 1/9 in the second, S is 100. If the shunt is disconnected entirely in the second test, S is 1000. It is safer to leave the high resistance in circuit in the second test to protect the galvanometer in case the insulation resistance is low. Therefore this resistance must be subtracted from the result to obtain the insulation of the machine itself.

By the above method it is possible to measure 100 megohms or even more. The wires and connections should be carefully arranged to avoid any possibility of contact or leakage which would spoil the test. If no deflection is obtained, place one finger on the frame and one on the binding post of the machine, which makes enough leakage to affect the galvanometer and show that the connections are right, thus proving that any poor insulation will be indicated, if it exists.

Fig. 77. Typical D'Arsonval Voltmeter
Courtesy of General Electric Company

Voltmeter Method. The voltmeter test for insulation resistance requires a sensitive high-resistance voltmeter. Take, for example, the 150-volt instrument, Fig. 77, which usually has about a 15,000-ohm resistance. (A certificate of the exact resistance is pasted inside each case.) Apply it to some circuit or battery and measure the voltage. This should be as high as possible, say 100 volts. The insulation resistance of the machine is then connected into the circuit, as indicated in Fig. 78. The deflection of the voltmeter is less than before, in proportion to the value of the insulation resistance. The insulation is then found by the equation:

$$\text{Insulation resistance} = \frac{DR}{d} - R$$

in which D is the first deflection; d is the second deflection; and R is the resistance of the voltmeter. If the circuit is 100 volts, D is 100; and if d, the deflection through the insulation resistance of

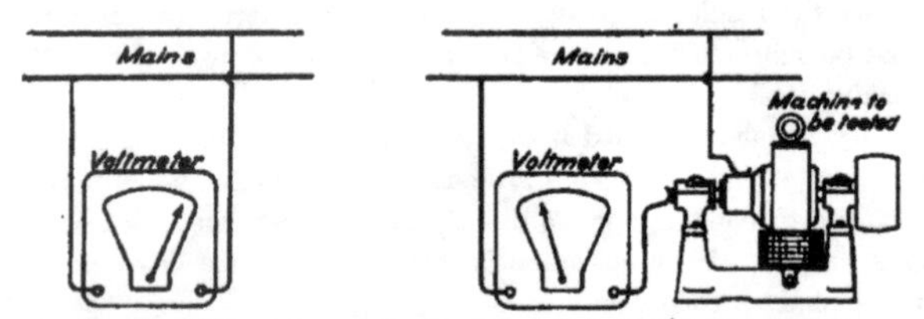

Fig. 78. Wiring Diagram for Insulation Test by Voltmeter Method

the machine, is 1 division, the insulation is 1,485,000 ohms. Permanent marks indicating amounts of insulation may be put on the voltmeter scale. When making measurements, the voltage should be the same as that employed in preparing this scale, say 115 volts. To calculate the scale use the formula

$$d = \frac{115\,R}{X + R}$$

in which X is the insulation resistance (1 megohm, ½ megohm, etc.); and d is the number of volts, opposite which the corresponding graduation is to be placed to form the new scale. This method does not test exceedingly high resistances; but if little or no deflection is obtained through the insulation resistance, it shows that the latter is at least several megohms—which is high enough for most practical purposes.

Magneto-Electric Bell. The ordinary magneto-electric bell may be used to test insulation by simply connecting one terminal to the binding-post of the machine and the other to the frame or shaft. A magneto bell is rated to ring with 10,000 to 30,000 ohms in series with it, and if it does not ring, it shows that the insulation is more than that amount. This limit is altogether too low for proper insulation in any case, and therefore this test is rough, and really shows only whether or not the insulation is very poor or the machine practically grounded.

The magneto is also used for "continuity" tests, to determine whether a circuit is complete, by simply connecting the two terminals of the magneto to those of the circuit. If the bell can be rung, it shows that the circuit is complete; if not, it indicates a break. An ordinary electric bell and cell of battery can be used in place of the magneto.

DIELECTRIC STRENGTH TESTS

The above methods of determining insulation are given in some detail, since they give a means of testing which can be applied almost any place. However, the ohmic resistance of any insulation is not of first importance. The dielectric strength, that is, the resistance to actual rupture by high voltage current, is considered as much more important. In making such tests, the high voltage should be applied between every circuit of the apparatus and every other circuit, and between every circuit and the material of the machine itself.

Test Voltages for Different Sized Machines. The rules of the American Institute of Electrical Engineers recommend that the proper voltage to be used for testing machines in general shall be twice the normal voltage of the circuit to which they are connected, plus 1000 volts. There are a few exceptions.

Conditions of Test. The machine should be tested with an alternating voltage of a virtual value (the ordinary a. c. voltmeter gives this value very closely), as given above. The frequency is not important on direct-current apparatus or on alternating-current apparatus of small capacity. However, on large a. c. machines the frequency should be the same as the normal frequency at which the machine is to operate.

Before being given high potential tests, all machines should be completely assembled, should be clean and in good running condition, and should be at a temperature corresponding to full load on the machine. The rules of the A. I. E. E. specify that the high-potential tests shall ordinarily be made in the factory of the manufacturer. A number of other rules are given in regard to "methods of testing", "methods of measuring the test voltage", and "apparatus for supplying the test voltage". These rules are followed almost universally in the United States today, as are the other rules of the A. I. E. E.

VOLTAGE

Voltmeter. An instrument used to measure the pressure or voltage of a circuit is called a voltmeter, Fig. 77. It is usually a galvanometer of practically constant resistance. The moving element carries a pointer which moves over a graduated scale. This scale is marked directly in volts or millivolts. A voltmeter should have a very high resistance in order that the indications may always be accurate and that the instrument may take as little current as possible. It should be shielded so that it will not be affected by stray fields due to magnets or currents flowing in conductors close to the instrument.

Test Method. The voltage of any machine or circuit is tested by merely connecting the two terminals of the voltmeter to the two terminals or conductors of the machine or circuit. To get the external voltage of a generator or motor, the voltmeter is usually applied to the two main terminals or brushes of the machine. This external voltage is what a generator supplies to the circuit. It is also called the *difference of potential*, or *terminal voltage*, and is the actual figure upon which calculations of the efficiency, capacity, etc., of any machine are based.

A generator for constant-potential circuits should, of course, give as nearly as possible a constant voltage. A plain shunt machine usually falls from 5 to 15 per cent in voltage when its current is varied from zero to full load. This is due to the i. r. drop caused by the resistance of the armature circuit, which, in turn, weakens the field current and magnetism. Armature reaction usually occurs also, and still further lowers the external voltage. This variation is undesirable, and is usually avoided by regulating the field magnetism (varying the resistance in the field circuit) or by the use of compound-wound generators. Compound-wound generators may be designed to give practically constant voltage from no load to full load, or they may be designed to rise any desired percentage of "over-compounding" from no load to full load.

CURRENT

Instruments for measuring current are called "ammeters". These instruments are built on the same general principle as a voltmeter, except that the main line current or a shunted part of it passes through the instrument.

Ammeter Always Connected in Series with Line. In testing the current of a generator or motor, it is necessary only to connect an ammeter of the proper range in series with the machine to be tested, so that the whole current passes through the instrument or its shunt. To test the current in the armature or the field alone, the ammeter is connected in series with the particular part. To avoid mistakes in the case of a shunt-wound generator, it is well to open the external circuit entirely in testing the current used in the field coils; for the same reason the brushes of a shunt motor should be raised before testing the current taken by the field.* In a constant-current, or series-wound, machine the same current flows through all parts as well as through the circuit; consequently the measurement of current is very simple.

If an ammeter cannot be had, current can be measured by inserting a known resistance in the circuit and measuring the difference of potential between its ends. The volts thus indicated, divided by the resistance in ohms, give the number of amperes flowing. If a known resistance is not at hand, the resistance of a part of the wire forming the circuit can be calculated from its diameter measured with a screw caliper or a wire gage, by referring to any of the tables of resistances of wires; or the resistance can be measured by a Wheatstone bridge, Fig. 74.

The above methods will seldom be found necessary in modern plants of any kind, since good portable ammeters are now so cheap as to be included in the equipment of nearly all such plants. In

* These instructions are to be followed when only one ammeter is to be had; otherwise one could be placed in the field circuit and another in the circuit from the starting box to the independent armature terminal.

many industrial plants, especially those using alternating current, it is the practice to put an ammeter on each motor circuit permanently. In central stations all machines and most feeder circuits are provided with ammeters for reading the line current.

Water Box for Testing Generators. In testing a generator at full load it is often difficult to find a means of disposing of the current generated. Any metallic resistance designed to absorb the entire output of a large generator would be out of the question because of the expense. If there is no way to absorb the power by connecting the machine to actually loaded distributing lines, a water box forms the cheapest means. The size of the box will, of

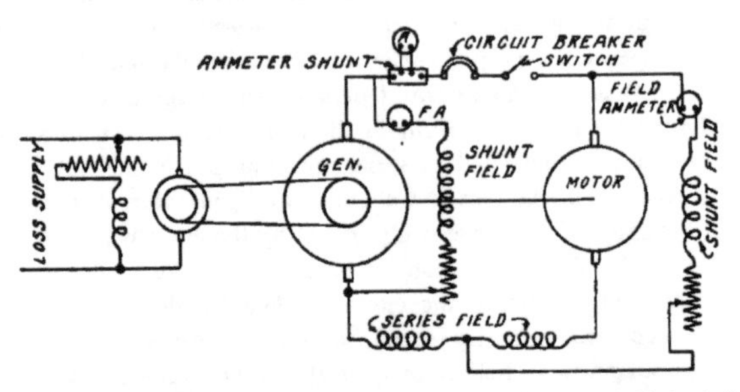

Fig. 79. Wiring Diagram for "Pump Back" Testing Method with Mechanical Loss Supply

course, depend upon the size of the machine to be tested. For machines of medium size a tub, barrel, or hogshead may be used. This is filled with water, and salt or common baking soda added until the proper current is obtained when the terminals are lowered into it. These terminals are usually two plates of iron to which the main conductors are attached. This arrangement allows the current to be varied by raising and lowering the plates, by adjusting the distance between plates, or by changing the strength of the solution. Alternating-current generators, when tested in this way, will require two, three, or four plates, depending on whether they are single-phase, three-phase, or two-phase.

"Pump Back" Method. In some cases, where duplicate machines are to be tested, another method is available. The two

MADISON RIVER POWER HOUSE NO. 2
Courtesy of Montana Power Company

machines are connected together both electrically and mechanically. Thus one machine, acting as a motor, drives the second machine mechanically; while the second machine, acting as a generator, drives the first machine electrically. Of course such a combination must be supplied with some other power to furnish the losses in the two machines. This can be done by belting another motor, operated from an outside source, to the two machines in test, Fig. 79.

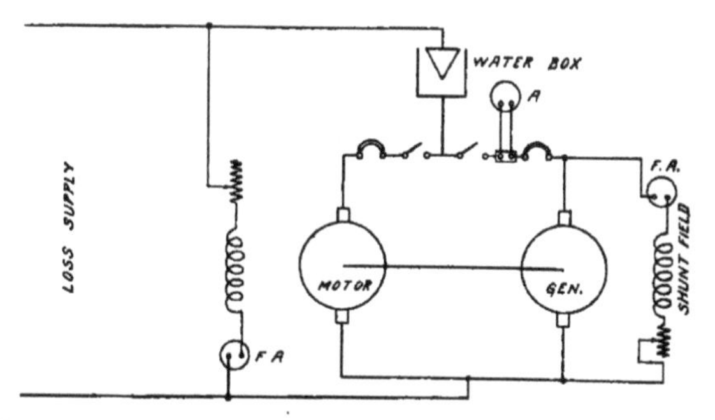

Fig. 80. Wiring Diagram for "Pump Back" Testing Method with Electrical Loss Supply

This auxiliary motor may be comparatively small since it supplies the losses only. With such an arrangement, one of the machines can be given full load and tested completely without the use of water boxes and without requiring a large amount of power to operate it.

In another of the "pump back" methods, there is an electrical loss supply instead of a mechanical loss supply. Fig. 80 shows the method of wiring the machines for such a test.

Similar methods may be employed for testing duplicate a. c. machines. Modification will have to be made, of course, to take care of the difference in the character of the current.

SPEED

Speed Counter. Speed is usually measured by the well-known speed counter, Fig. 81, consisting of a small spindle which turns a wheel one tooth each time it revolves. The point of the spindle is held against the center of the shaft of the generator or motor for a certain time, say one minute or one-half minute, and the number of revolutions is read off from the position of the wheel.

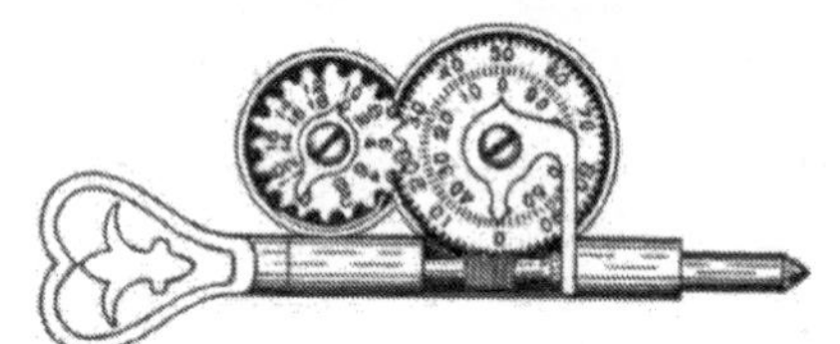

Fig. 81. Typical Speed Counter

Tachometer. Another instrument for testing the number of revolutions per minute is the tachometer. The stationary form of this instrument is shown in Fig. 82. It must be belted by a string, tape, or light leather belt to the machine, the speed of which is to be tested. If the sizes of the pulleys are not the same, their speeds are inversely proportional to their diameters. The portable form of this instrument, Fig. 83, is applied directly to the end of the shaft of the machine, like the speed counter, a sliding gear shifter providing the required range of speeds. These instruments possess the great advantage over the speed counter that they instantly point on the dial to the proper speed, and they do not require to be timed for a certain period.

Fig. 82. Stationary Form of Tachometer
Courtesy of James G. Biddle, Philadelphia

Electric Tachometer. Another instrument which gives a direct reading of revolutions per minute is the electric tachometer. This instrument consists of a small generator, which is driven by the revolving shaft, and a voltmeter connected across the terminals of this generator. The design is such that the voltage of the generator is proportional to its speed. The voltmeter is calibrated in revolutions per minute and is, therefore, a direct-reading tachometer.

A simple way to test the speed of a belted machine in revolutions per minute is to make a large black or white mark on the belt and note how many times the mark passes per minute; the length of the belt divided by the circumference of the pulley gives the number of revolutions of the pulley for each time the mark passes. The number of revolutions of the pulley to one of the belt can also be easily determined by slowly turning the pulley or pulling the belt until the latter makes one complete trip around, at the same time counting the revolutions of the pulley. If the machine has no belt, it can be supplied with one temporarily for the purpose of the test, a piece of tape with a knot or an ink mark being sufficient. Care should be taken in all these tests of speed with belts not to allow any slip; for example, in the case of the tape belt just referred to, this belt should pass around the pulley of the machine and some light wheel of wood or metal which turns so easily as not to cause any slip of the belt on the pulley of the machine. The belt should not have much elasticity like a rubber band, as it would give an incorrect result.

Fig. 83. Hand Tachometer
Courtesy of James G. Biddle

TORQUE

Prony Brake. Torque, or pull, is measured in the case of a motor by the use of a Prony or strap brake. The former consists of a lever $L\,L$ of wood, clamped on the pulley of the machine to be tested, as indicated in Fig. 84. The pressure of the screws $S\,S$ is then adjusted by the wing-nuts until the friction of the clamp on the pulley is sufficient to cause the motor to take a given current, and the speed is then noted. Usually, the maximum torque or pull is the most important to test; and this is obtained in the case of a constant-potential motor by tightening the screws S S until the motor draws its full current, as indicated by an ammeter. The full-load current is usually marked on the name plate of the motor. If it is not known, the following formulas may be used:

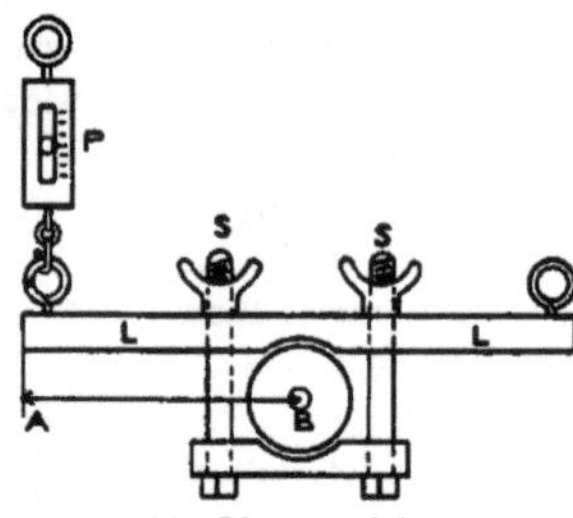

Fig. 84. Diagram of Prony Brake

For d. c. motors:

$$\text{Current} = \frac{\text{h.p.} \times 746}{\text{Volts} \times \text{Efficiency}}$$

For single phase a. c. motors:

$$\text{Current} = \frac{\text{h.p.} \times 746}{\text{Volts} \times \text{Efficiency} \times \text{Power Factor}}$$

For three-phase a. c. motors:

$$\text{Current} = \frac{\text{h.p.} \times 746}{\text{Volts} \times \text{Efficiency} \times \text{Power Factor} \times 1.73}$$

For two-phase a. c. motors:

$$\text{Current} = \frac{\text{h.p.} \times 746}{\text{Volts} \times \text{Efficiency} \times \text{Power Factor} \times 2}$$

The efficiency and the power factor in these formulas should always be expressed as decimals. When the rating is given in kw., instead of h. p., this rating multiplied by 1000 should be substituted instead of h. p. × 746.

The torque or pull is measured by known weights or, more conveniently, by a spring balance *P*. If desired, the test may also be made at three-quarters, one-half, or any other fraction of the full-load current.

The torque in foot-pounds, which should be obtained, can also be calculated from the power at which the machine is rated, by the formula:

$$\text{Torque} = \frac{\text{h.p.} \times 33{,}000}{6.28 \times S}$$

or

$$P = \frac{\text{h.p.} \times 33{,}000}{2\pi r \times S}$$

in which h. p. is the horsepower of the machine at full load, and S is the speed of the machine in revolutions per minute at full load. Torque is given at unit radius, commonly pounds at one foot. The pull at any other radius is converted into torque by multiplying by the radius. One h. p. produced at a speed of 1000 revolutions requires a pull of 5.25 pounds at the end of a 1-foot lever; at 500 revolutions, twice as much; at 2000 revolutions, half as much; and so on. If the lever is 4 feet, the pull is one-fourth as much, etc.

Torque of a Generator. The torque of a generator, that is, the power required to drive it, is very conveniently determined by operating it as a motor and testing it by the friction brake, as described above, the torque of a generator being practically equal to that of a motor under similar conditions.

POWER

Electrical Power. The electrical power of a generator or motor is found by testing the voltage and the current at the terminals of the machine, as already described, and multiplying the two together, which gives the electrical power of the machine in watts.* Watts are converted into horsepower by dividing by 746, and into kilowatts by dividing by 1000.

Mechanical Power. The mechanical power of a generator or motor, that is, the power required for or developed by it, is found by multiplying its pull by its speed and by the circumference on which the pull is measured, and dividing by 33,000. That is,

$$\text{Horsepower} = \frac{P \times S \times 6.28 \times R}{33{,}000}$$

* In testing an alternating-current machine, a wattmeter should be employed instead of a voltmeter and an ammeter, as explained later.

in which P is the pull in pounds; S is the speed in revolutions per minute; and R is the radius in feet at which P is measured.

Efficiency

Analysis of Definition. The efficiency is determined in the case of a generator by dividing the electrical power generated by it by the mechanical power required to drive it; that is,

$$\text{Efficiency of generator} = \frac{\text{Electrical power}}{\text{Mechanical power}}$$

The efficiency of a motor is the mechanical power developed by it, divided by the electrical power supplied to it; that is,

$$\text{Efficiency of motor} = \frac{\text{Mechanical power}}{\text{Electrical power}}$$

Efficiency of D.C. Generators. *Outline of Method.* In testing the efficiency of d. c. generators there are several methods which can be followed. One large manufacturer has adopted the "measurement of losses" method, and it is probably one of the best for machines of all sizes. In this method the following quantities are measured:

Voltage of the line
Current in the line
Current in the shunt field
Current in the armature
Current in the series field
Current in the series field shunt
Resistance of the brush contact
Resistance of the shunt field (hot)
Resistance of the armature (hot)
Resistance of the series field (hot)
Resistance of the series field shunt (hot)
Core loss
Brush friction loss
Bearing friction loss

Core-Loss and Friction Test. These last three items are determined by a separate test in the following manner: Drive the generator by a small motor, say 10 per cent of the capacity of the generator. Belt drive is usually satisfactory but, if great accuracy is necessary, the motor should be direct-connected to the generator. The motor should not have to carry more than 50 per cent of its full-load rating with maximum field strength on the generator. The motor should be run at its normal speed and field strength as nearly as possible. The resistance of the motor armature should be known. The field strength must be kept *constant* at about normal value. The speed of the motor can be varied by changing the voltage across its armature. The motor and the generator should be run for some time to allow all friction conditions to become constant. Readings should be obtained as follows:

Input to the motor with all brushes down and no field on the generator.
Input to the motor with all brushes raised and no field on the generator.

(The difference in the above gives the loss in brush friction.)

Input to the motor with various values of field current in the generator from zero up to such a value as will give voltage considerably above normal.

Input to the motor while running free with the same field strength as used during the balance of the test.

The speed must be held absolutely constant and no readings taken when the speed is changing in either direction. It is readily seen that by correcting the input to the motor to allow for the I^2R loss in the armature, and for the input to the motor with no field on the generator, we can get the core loss at the different field strengths.

Final Efficiency. We can now add the I^2R losses in the different sections of the electric circuits to the core loss and the friction losses, and obtain the total loss in the generator. The total input then becomes:

$$W_1 = W_O + L$$

$$E = \frac{W_O}{W_1} = \frac{W_O}{W_O + L}$$

in which W_1 is watts input; W_0 is watts output; L is total loss in watts; and E is efficiency.

These losses might have values somewhat as given below for a 20 kw. generator:

Core loss .... about 3%
Bearing friction and windage.................. about 2½%
Brush friction .... about ½%
I^2R losses................................. about 2 to 4%

This would mean an efficiency of about 91 or 92 per cent. The losses given above can be further divided, but such division is not usually necessary. It is to be noted that there are in practice additional losses called "load losses", very difficult to measure. These are usually about 1 or 2 per cent. The method described above can be used for any direct-current machine.

Efficiency of A. C. Generators. For a.c. machines a similar method is used, but in addition to the open-circuit core loss, the core loss is measured with the armature short-circuited and sufficient field to give normal armature current. The total losses obtained are added to the output to obtain the input the same as for d.c. machines, and the efficiencies are obtained in the same way.

Efficiency of Motor-Generators and Converters. The efficiency of a motor-generator or ordinary converter is very easily determined by simply measuring the input and output in watts (by wattmeters or by ammeters and voltmeters for direct currents), and dividing the latter by the former. These electrical methods of testing are preferable to mechanical, for the reason that the volts and the amperes can be easily and accurately measured, and their product gives the power in watts.* Mechanical measurements of power by dynamometer or other means are more difficult, and not so accurate, but brake tests give good practical results.

Measurement of Power in A. C. Circuits

In circuits carrying alternating currents and having some inductive load either in the form of motors or arc lamps or a partly loaded transformer, etc., the ordinary method of determining the power, by voltmeter and ammeter measurements, is not applicable, as the current is seldom in phase with the e.m.f. and, therefore, the product *volts* × *amperes* is not the true power.

Indicating Wattmeter Method. There are several means for determining the true power of an a.c. circuit, the simplest being an indicating wattmeter. A wattmeter is an electrodynamometer

* When alternating-current machinery is being tested, use wattmeters.

provided with two coils, a fixed one of coarse wire, the other movable and of fine wire. This movable coil is connected in series with a large noninductive resistance, so that the time constant of the fine-wire circuit is extremely small, and hence its impedance is practically equal to its resistance; the current in, and the resulting field of, the fine-wire coil will under these conditions be practically in phase with the potential difference across its terminals. The field produced by the coarse-wire coil is directly proportional to the current flowing through it at any instant. Hence, the couple acting on the fine-wire coil is proportional at a given instant to the product of these two fields; so that the reading of the instrument, which depends on the mean value of the couple, will be proportional to the mean power, and, by providing the instrument with the proper scale, it will read directly in watts.

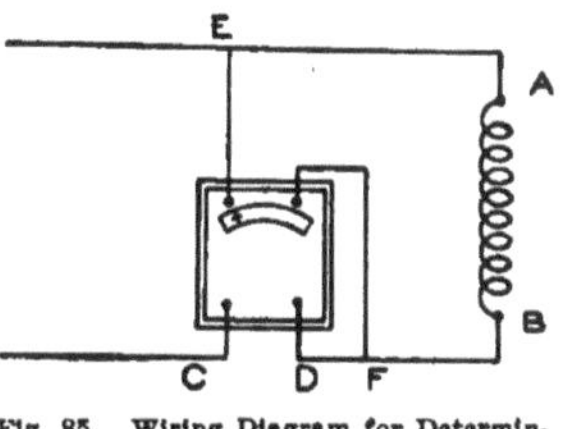

Fig. 85. Wiring Diagram for Determining Input by Indicating Wattmeter Method in Single-Phase Circuit

Single-Phase Circuit. In Fig. 85, $A\,B$ represents an inductive load—say of a single-phase motor—of which the power input is to

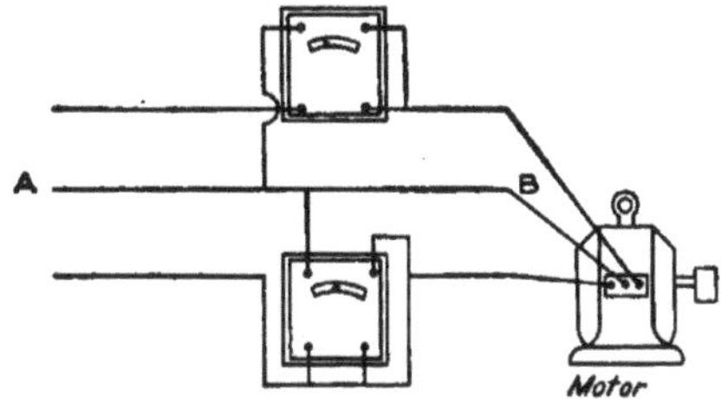

Fig. 86. Wiring Diagram for Testing Power Input in Two-Phase Circuit

be determined; $C\,D$, the terminals of the thick-wire coil (current-coil) of the wattmeter; and $E\,F$, the voltage- or pressure-coil terminals. When connected as above indicated, the wattmeter indicates directly in watts the power supplied.

Two-Phase Circuit. In the case of a two-phase system, where the two circuits are independent, the power may be measured by placing a wattmeter in each phase, as shown in Fig. 85, and adding the two readings. If the motor be connected up as shown in

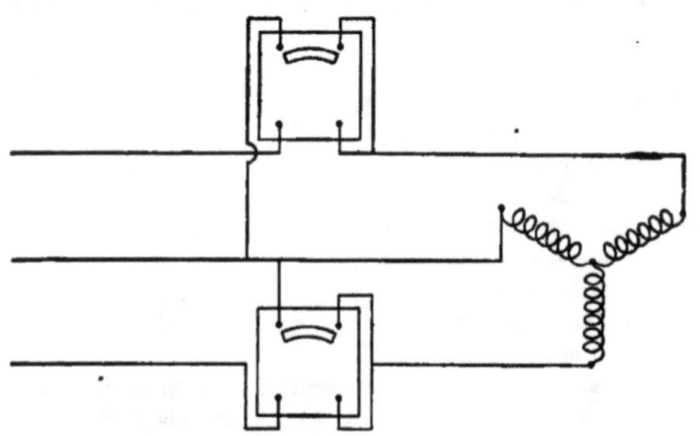

Fig. 87. Power Input Diagram for Three-Phase Circuit, Star-Connected

Fig. 86,* where $A B$ forms a common return, the wattmeters are placed as indicated, *care being taken to place the current-coils in*

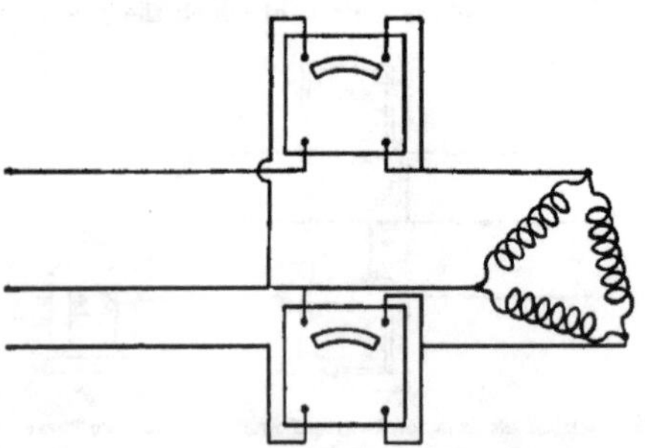

Fig. 88. Power Input Diagram for Three-Phase Circuit, Delta-Connected

the outside mains; and the power supplied is equal to the sum of the two wattmeter readings.

* This form of connection is possible only when the generator has two independent windings, one for each phase.

Three-Phase Circuit. The power of a balanced or unbalanced three-phase system can be determined by the use of two wattmeters connected as shown in Figs. 87 and 88. The current-carrying coils are placed in series with two of the wires and the pressure coils, respectively, connected between these two mains and the third wire. The *algebraic* sum of these two wattmeter readings gives the true power supplied. When the power factor of the system is less than .5, one of the wattmeters will read negatively. It is sometimes difficult to determine whether the smaller readings are negative or not. If in doubt, give the wattmeter a separate load of incandescent lamps, and make the connections such that both instruments deflect positively; then reconnect them to the load to be measured. If the terminals

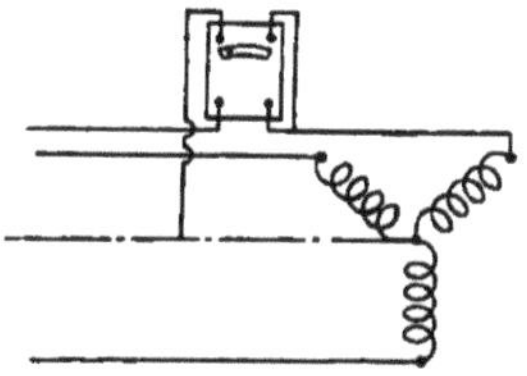

Fig. 89. Power Input Diagram for a Four-Wire Three-Phase Circuit with One Wattmeter

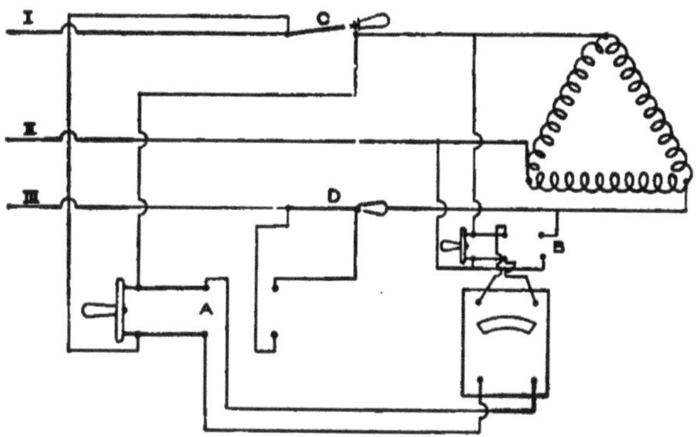

Fig. 90. Power Input Diagram for a Three-Phase Three-Wire Circuit with One Wattmeter

of one instrument have to be reversed, the readings of that wattmeter are negative.

Balanced Four-Wire Three-Phase Circuit. To measure the power of a balanced four-wire three-phase system, one wattmeter may be connected as shown in Fig. 89, and the wattmeter reading multiplied by 3. Usually, however, a four-wire three-phase system is unbalanced; and to determine the power supplied under this condition, three wattmeters should be employed, one for each phase

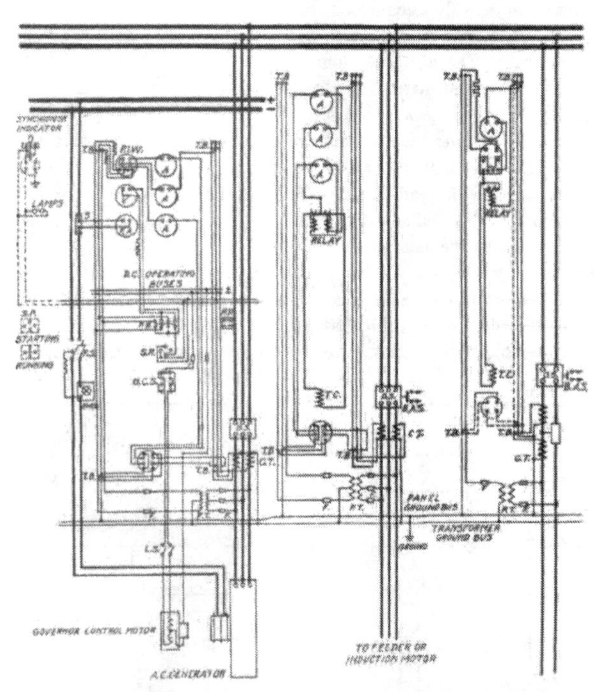

Fig. 91. Wiring Diagram for Switchboard Equipment of Standard Central Station

the power supplied being equal to the algebraic sum of all three readings.

It is obvious that in any of the above instances one wattmeter could be employed, provided the necessary switches are used. Assuming, for example, the three-phase three-wire case, one wattmeter would require switch connections as shown in Fig. 90. *A* is a double-pole switch which, when thrown to the left, places the current coil of the wattmeter in series with the conductor of No. I, and, when thrown to the right, places it in series with No. III. Similarly, switch *B* changes the pressure terminals from between machines I and II to machines III and II; while switches *C* and *D* are short-circuiting switches. One of these switches is closed previous to removing the current coil from one phase to the other, and the other one is opened after the coil is in the position indicated in the diagram.

In general practice, power in polyphase circuits is measured by polyphase wattmeters. Such an instrument is merely a combination of two or three single-phase instruments. Thus for two-phase non-interconnected systems, two single-phase elements are mounted together with the two moving elements mounted on the same shaft and carrying one pointer. Thus the movement of the pointer is caused by the combined torque of the two elements and the indication on the scale is equal to the algebraic sum of the two.

Three elements could be combined in the same way for three-phase four-wire circuits. In practice a third current element is customarily added to the usual two of a three-wire instrument. This third coil is divided, each half of it being placed so as to act with one of the two potential coils. This arrangement does not give an absolutely correct indication, but the error is so small that it does not throw the indications outside the limits of commercial accuracy.

Instrument Transformers. In a great many cases in modern practice it is necessary to use instrument transformers. Where the current involved is more than 200 or 300 amperes, or in some cases even less, the instruments can not be made to carry the full amount and current transformers are used to step the current down. The secondary current is usually 5 amperes. Where the voltage is more than 1100, both current and potential transformers are usually used, even for small currents, since it is not safe to bring high voltage directly into the instrument. Fig. 91 shows the wiring diagram of a standard central station switchboard in which the instruments are operated from current and potential transformers.

LOCALIZATION AND REMEDY OF TROUBLES

General Plan. The promptness and ease with which any accident or difficulty with electrical machinery can be dealt with will always have much to do with the success of a plant. A list of troubles, symptoms, and remedies for the various types and sizes of dynamos and motors in common use has been prepared to facilitate the detection and elimination of such difficulties.

It is evident that the subject is somewhat complicated and difficult to handle in a general way, since so much depends upon the particular conditions in any given case, every one of which must be included in the "Table of Troubles" in such a way as to distinguish it from all the others. Nevertheless, it is remarkable how much can be covered by a systematic statement of the matter, and nearly all cases of trouble most likely to occur are covered by the table, so that the detection and remedy of the defect will result from a proper application of the rules given.

It frequently happens that a trifling oversight, such as allowing a wire to slip out of a terminal, will cause as much annoyance and delay in the use of electrical machinery as the most serious accident. Other troubles, equally simple but not so easily detected, are of frequent occurrence.

The rules are made, as far as possible, self-explanatory; but a statement of the general plan followed in its most important features will facilitate the understanding and use of the table.

USE OF TABLE OF TROUBLES

In the use of the Table of Troubles the principal object should be to separate clearly the various causes and effects from one another. A careful and thorough examination should first be made; and, as far as possible, one should be perfectly sure of the facts, rather than attempt to guess what they are and jump at conclusions. Of course, one should take general precautions and preventive measures before any troubles occur, if possible, rather than wait until a difficulty has arisen. For example, one should see that the machine is not overloaded or running at too high voltage, and should make sure that there is oil in the bearings. Neglect and carelessness with any machine are usually and deservedly followed by accidents of some sort. It is usually wise to stop the machine, if possible, when any trouble manifests itself, even though it does not seem to be very serious. It is often practically impossible to shut down altogether; therefore spare apparatus should always be ready. The continued use of defective machinery is a common but very objectionable practice.

Classification of Troubles. The general plan of the table is to divide all troubles that may occur to generators or motors into ten classes, the headings of which are the ten most important and obvious bad effects produced in these machines, viz:

Table of Troubles

I. Sparking at Commutator
II. Heating of Commutator and Brushes
III. Heating of Armature
IV. Heating of Field Magnets
V. Heating of Bearings
VI. Noisy Operation
VII. Speed Not Right
VIII. Motor Stops or Fails to Start
IX. Dynamo Fails to Generate
X. Voltage Not Right

Any one of these general effects is evident, even to the casual observer, and still more so to any person making a careful examination; hence nine-tenths of the possible cases can be eliminated immediately.

The next step is to find out which particular one of the eight or ten causes in this class is responsible for the trouble. This requires more careful examination, but nevertheless can be done with comparative ease in most cases. One cause may produce two effects, and, *vice versa,* one effect may be produced by two causes; but the table is arranged to cover this fact as far as possible. In a complicated or difficult case, it is well to read through the entire table and note what causes can possibly apply. Generally, there will not be more than two or three; and the particular one can be picked out by following the directions, which show how each case may be distinguished from any other.

I. SPARKING AT COMMUTATOR

This is one of the most common of troubles, being often quite serious because it burns and cuts the commutator and brushes, at the same time producing heat that may spread to and injure the armature or bearings. Any machine having a commutator, including practically all direct-current and some alternating-current machines, is liable to have this trouble. The latter usually have continuous collecting rings not likely to spark; but self-exciting or composite-wound alternators, rotary converters, and some alternating-current motors have supplementary direct-current commutators. A certain amount of sparking occurs normally in most constant-current dynamos for arc lighting, where it is not very objectionable, since the machines are designed to stand it and the current is small.

Cause 1. Armature carrying too much current. This is due to (a) overload (for example, too many lamps, motors, etc., fed by generator, or too much mechanical work done by motor; a short circuit, or ground on the line, may also have the effect of overloading a generator); (b) excessive voltage on a constant-potential circuit, (or excessive amperes on a constant-current circuit). In the case of a motor, any friction, such as armature striking pole pieces, or shaft not turning freely, may have the same effect as overload.

Symptom. Whole armature becomes overheated, and belt (if any) becomes very tight on tension side, sometimes squeaking because of slipping on pulley. Overload due to friction is detected by stopping the machine, and then turning it slowly. (See V and VI, 2.)

Remedy. (a) Reduce the load, or eliminate the short circuit or ground on the line; (b) decrease size of driving pulley; or (c) increase size of driven pulley; (d) decrease magnetic strength of field in the case of a generator, or increase it in the case of a motor. If excess of current can not satisfactorily be overcome in any of the above ways, it will be necessary to change the machine or its winding. Overload due to friction is eliminated as described under V and VI, 2.

If the starting or regulating rheostat of a motor has too little resistance it will cause the motor to start too suddenly and to spark badly at first. The only remedy is more resistance.

Cause 2. Brushes not set at proper commutating point.

Symptom. Sparking at the brushes.

Remedy. Brushes should be carefully set on the proper neutral point on the commutator. For non-commutating-pole machines this is either ahead of or behind (depending whether the machine is a generator or a motor) a radial line on the commutator usually passing through the center of the pole piece. This point is found by shifting the brushes from the geometrical or "mechanical" neutral so that at no load and normal voltage a slight spark, the size of a pin point, shows at the edge of the brush. This point will usually be the proper full-load running position of the brushes. While the usual position of the brushes is about opposite the center of the pole piece, in some machines it may be opposite the space between adjacent pole pieces, depending on the winding of the armature. If the brushes are set exactly wrong, this will cause a generator to fail to generate, and a motor to fail to start, and will blow the fuse or open the breaker in the latter case. (See IX, 6.)

The proper brush position for commutating-pole generators coincides very closely with the mechanical neutral, and is best determined thus: Remove one brush from a brush holder and in its place insert a fiber brush of similar dimensions. Through this drill two holes, about the size of the lead in a pencil, at such a distance apart that the holes are respectively over the centers of two adjacent commutator-bars. Insert in these holes the lead from a pencil so that the ends of the lead ride lightly on the commutator. To the outer ends connect a low-reading voltmeter, and with normal voltage on the generator move the brush-holder yoke until the voltmeter shows a zero reading. This will be the correct brush position, or "electrical neutral".

In the case of a commutating-pole motor, the electrical neutral is determined by shifting the brushes until the same speed is obtained in either direction, with equal value of field current and equal values of voltage.

In the case of all d.c. generators and motors it is of prime importance that the brushes be placed parallel to the commutator-bars, and the various sets of brushes spaced exactly the same distance apart.

Cause 3. Commutator rough, eccentric. The commutator may be rough or eccentric; may have one or more *high bars* projecting above the others; one or more *low bars*, or *projecting mica;* or may have as many *low black spots on the commutator* as there are pairs of poles on the machine. Any one of these may cause the brushes to vibrate or to be actually thrown out of contact with the commutator.

Symptom. Note whether there is a glaze or polish on the commutator which shows smooth operation; touch the revolving commutator with the point of a pencil and any roughness or low or high spots can be detected. In the case of an eccentric commutator careful examination shows a rise or fall of the brushes when the commutator turns slowly, or a chattering of the brush when it is running fast.

Remedy. First, go all over the armature, commutator, and equalizer connections, and test for high resistance joints. If any are found they should at once be soldered. (This applies also to rotary converters.) The proper brush setting and spacing should be checked, also the air gap and magnetic joints on the machine. If the commutator is not in very bad shape it may be ground down with a piece of fine sandstone, or sand or carborundum paper (not emery). This should be fitted to a wood block cut to the same curvature as the commutator.

If the commutator is in very bad shape or is eccentric, the armature should be taken out and put in a lathe and the commutator turned. Large machines are fitted with a slide rest attachment for turning the commutator without removing it.

For turning the commutator, a diamond point tool should be used. Only a fine cut should be taken off each time in order to avoid catching in, or chattering on, the copper bars, which are very tough. The surface is then finished with either a sandstone or fine sandpaper.

In order that the commutator may wear smooth and work well, the armature shaft should move freely back and forth about $\frac{1}{8}$ or $\frac{3}{16}$ of an inch in its bearings while running.

A commutator should have a glaze of a brown color. A very

bright or scraped appearance does not indicate the best condition. A light grade of engine oil used sparingly on the commutator is beneficial.

Cause 4. Brushes make poor contact with commutator.

Symptom. Close examination shows that brushes touch only at one corner, or only in front or behind, or there is dirt on surface of contact. Sometimes, owing to the presence of too much oil or from other cause, the brushes and commutator become very dirty, and covered with smut. They should then be carefully cleaned by wiping with oily rag or benzine, or by other means.

Occasionally a "glass-hard" carbon brush is met with. It is incapable of wearing to a good seat or contact, and will touch at only one or two points. Some carbon brushes are of abnormally high resistance, so that they do not make good contact. In such cases new brushes should be substituted.

Fig. 92. Typical Jig for Carbon Brushes

Remedy. Carefully fit, adjust, or clean brushes until they rest evenly on the commutator, with considerable surface of contact and with sure but not too heavy pressure. Copper brushes require a regular brush jig, Fig. 92. Carbon brushes can be fitted perfectly by drawing a strip of sandpaper back and forth between them and the commutator while they are pressing down. A band of sandpaper may be pasted or tied around the commutator, and the armature then slowly revolved by hand or by power while the brushes are pressed upon it.

Cause 5. Short-circuited or reversed coil or coils in armature.

Symptom. A motor will draw excessive current, even when running free without load. A generator will require considerable power, even without any load. For reversed coil, see III, 5.

The short-circuited coil is heated much more than the others, and is liable to be burnt out entirely; therefore the machine should be stopped immediately. If necessary to run machine in order to locate the trouble, one or two minutes is long enough; but this may be repeated until the short-circuited coil is found by feeling the armature all over.

An iron screw driver or other tool held between the poles near the revolving armature vibrates very perceptibly as the short-circuited coil passes. Almost any armature will cause a slight but rapid vibration of a piece of iron held near it; but a short circuit produces a much stronger effect only *once* per revolution. Care should be taken not to let the piece of iron be drawn in and jam the armature.

The current pulsates and torque is unequal at different parts of a revolution, these symptoms being particularly noticeable when several coils are short-circuited or reversed, and the armature is slowly turned. If a large portion of the armature is short-circuited, the heating is distributed and is harder to locate. In this case a motor runs very slowly, giving little power but having full field magnetism. A short-circuited coil can also be detected by the drop-of-potential method. For generators, see IX, 3.

Remedy. A short circuit is often caused by a piece of solder, copper, or other metal getting between the commutator-bars or their connections with the armature; and sometimes the insulation between or at the ends of these bars is bridged over by a particle of metal. In any such case the trouble is easily found and corrected. If, however, the short circuit is in the coil itself, the only effective remedy is to rewind the coil.

One or more "grounds" in the armature may produce effects similar to those arising from a short circuit. (See Cause 7.)

Cause 6. Broken circuit in armature.

Symptom. Commutator flashes violently while running, and commutator-bar nearest the break is badly burnt; but in this case no particular armature coil will be heated as in the last case; and the flashing will be very much worse, even when turning slowly. This trouble, which might be confounded with a bad case of "high bar" in commutator (Cause 3), is distinguished therefrom by slowly turning the armature, when violent flashing will continue if circuit is broken; but not with high bar unless it is very bad, in which case it is easily felt or seen. A very bad contact has almost the same effect as a break in the circuit.

Remedy. A break or bad contact can be located by the

"drop" method (p. 88) or by a continuity test (p. 93). The trouble is often found where the armature wires connect with the commutator, and not in the coil itself; the break may be repaired or the loose wire fastened or reconnected. If the trouble is due to a broken commutator connection and can not be fixed, the disconnected bar may be temporarily connected to the next by solder. If the break is in the coil itself, rewinding is generally the only cure. The trouble may be remedied temporarily by connecting together by wire or solder the two commutator-bars or coil-terminals between which the break exists. It is only in an emergency that armature coils should be cut out or commutator-bars connected together, or other makeshifts resorted to; but it sometimes avoids a very undesirable shut-down. A very rough but quick and simple way to connect two commutator-bars is to hammer or otherwise force the coppers together across the mica insulation at the end of the commutator. This should be avoided if possible; but if it has to be done in an emergency, the crushed material can afterwards be picked out and the injury smoothed over. In carrying out any of these methods, great care should be taken not to short-circuit any other armature coil, which would cause sparking (Cause 5).

Cause 7. Ground in armature.

Symptom. Two "grounds" (accidental connections between the conductors on the armature and its iron core or the shaft or spider) would have practically the same effect as a short circuit (Cause 5), and would be treated in the same way. A single ground would have little or no effect, provided the circuit is not intentionally or accidentally grounded at some other point. On an electric-railway ("trolley") or other circuit employing the earth as a return conductor, one or more grounds in the armature would allow the current to pass directly through them, and would cause the motor to spark and have a variable torque at different parts of a revolution.

Remedy. A ground can be detected by testing with a magneto bell (p. 92). It can also be located by the drop-of-potential method (p. 88). Another way to locate it is to wrap a wire around the commutator so as to make connection with all of the bars, and then connect a source of current to this wire and to the armature core (by pressing a wire upon the latter). The current will then flow from the armature conductors through the ground connection to the core, and the magnetic effect of the armature winding will be localized at the point where the ground is. This point is then found by the indications of a compass needle when slowly moved around the surface of the armature. The current may be obtained from a storage battery or from the circuit, but should be regulated by lamps or other resistance so as not to exceed the normal armature current. Sometimes the ground may be in a place where it can be corrected without much trouble, but usually the particular coil and often others must be rewound.

Cause 8. Weak field magnetism.

Symptom. Pole pieces not strongly magnetic when tested with a piece of iron. Point of least sparking shifted considerably from normal position owing to relatively strong distorting effect of armature magnetism. Speed of a shunt motor high.* A generator fails to generate the full e. m. f. or current.

If one field coil is reversed and opposed to the others, it will weaken the field magnetism and cause bad sparking. This may be detected by examining the field coils to see if they are all connected in the right way, or by testing with a compass needle. (See IX, 4.) The series-coil of a compound-wound generator or motor is often connected wrongly, and will have an opposing effect; that is, will reduce the voltage of the former or raise the speed of the latter with increase of load.

Remedy. A break, short circuit, or ground, if external and therefore accessible, is easily repaired. If not accessible, the only remedy is to rewind or replace the faulty coil. A shunt motor will show dangerous sparking if the armature is connected before the field, in starting. In this case the starting-box connections should be changed so that the field is connected before the armature.

If the voltage is too low on a circuit, it may cause sparking in a shunt motor; and if the voltage cannot be raised, the resistance of the field circuit should be reduced by unwinding a few layers of wire or by substituting other coils. (See VII, VIII, IX, and X.)

* Under some circumstances a shunt motor with a very weak field, or with no field, will run slow, stop, or run backward. The usual and safest assumption is that the speed will increase as the field is weakened, and that with no field the motor may attain such a speed that it will destroy itself. *Never, under any circumstances, should the field circuit be opened when voltage is applied to the terminals.*

Cause 9. Vibration of machine.

Symptom. Considerable vibration is felt when the hand is placed upon the machine, and sparking decreases if the vibration is reduced.

Remedy. The vibration is usually due to an imperfectly balanced armature or pulley (see VI, 1), to a bad belt (see VI, 6), or to unsteady foundations; and the remedies described for these troubles should be applied.

Any considerable vibration is likely to produce sparking, of which it is a common cause. This sparking can be reduced by increasing the pressure of the brushes on the commutator, but the vibration itself should be overcome.

Cause 10. Chatter of brushes. The commutator sometimes becomes sticky when carbon brushes are used, causing friction, which throws the brushes into rapid vibration as the commutator revolves, similar to the action of a violin bow.

Symptom. Slight tingling or jarring is felt in brushes.

Remedy. Clean commutator, and oil slightly.

Cause 11. Flying break in armature conductor.

Symptom. No break found by test with armature standing still, but break shown by flashing at brushes when running, being usually due to centrifugal force.

Remedy. Tighten connections to commutator, or repair broken wire, etc.

II TO V. EXCESSIVE HEATING IN GENERATOR OR MOTOR

General Instructions. The degree of heat that is injurious or objectionable in a generator or motor is determined, as a rough test, by feeling the parts. If the heat is bearable to the hand, it is entirely harmless; but if unbearable, the safe limit of temperature has been approached or passed, and the heat should be reduced in some of the ways that are indicated below. In testing with the hand, allowance should be made for the fact that bare metal feels much hotter than cotton at the same temperature. The back of the hand is more sensitive than the palm for this test. If the heat has become so great as to produce an odor or smoke, the safe limit has been far exceeded, and the current should be shut off immediately and the machine stopped, as this indicates a serious trouble, such as a short-circuited coil or tight bearing. The machine should not again be started until the cause of the trouble has been found and positively overcome. Of course, neither water nor ice should ever be used to cool electrical machinery, except possibly the bearings of large machines at points where they can be applied *without danger of wetting the other parts.*

Feeling for heat will serve as a rough test to detect excessive temperatures or in emergencies; but, of course, the sensitiveness of the hand varies, and it makes a great difference whether the surface is a good or bad conductor of heat. The proper and reliable methods for determining rise in temperature in an operating machine are given on page 75.

It is very important, in all cases of heating, to locate the source of heat in the exact part in which it is produced. It is a common mistake to suppose that any part of the machine that is found to be hot is the seat of the trouble. A hot bearing may cause the armature or commutator to heat, or *vice versa.* In every case all parts of the machine should be tried to find which is the hottest, since heat generated in one part is rapidly diffused throughout the machine. It is better to make observations for heating by starting the whole machine when cool, which is done by letting it stand for several hours.

II. Heating of Commutator and Brushes

Cause 1. Heat spread from another part of machine.

Symptom. Start with the machine cool, and run for a short time, so that heat will not have time to spread. The real seat of trouble is the part that heats first.

Remedy. See III, IV, and V.

Cause 2. Sparking. Any of the causes of sparking will cause heating, which may be slight or serious.

Symptom and Remedy. See I.

Cause 3. Tendency to spark, or slight sparking hardly visible. Sometimes before sparking appears, serious heating is produced by the causes of sparking, such as the short-circuiting of the coils as their commutator bars pass under the brushes.

Symptom. Fine sparks may be found by sighting in exact

line with the surface of contact between the commutator and brushes.

Remedy. Reduced by applying the principal remedies for sparking, such as slightly shifting rocker-arm. (See "Sparking at Commutator", I.) Apply the remedies with extra care. This incipient sparking may be due to incorrect design, which can be corrected only by reconstruction; for example, it may be due to insufficient field strength, and this can be cured by increasing the ampere turns of field winding.

Cause 4. Overheated commutator will decompose carbon brush. The effect is to cover commutator with a black film, which offers resistance and aggravates the heat.

Symptom. Commutator covered with dark coating; commutator, brushes, and holders show marks of abnormal heat.

Remedy. Commutator and brushes should be carefully cleaned, and the latter adjusted to make good contact at the proper points.

Cause 5. Bad connections in brush holder, cable, etc.

Symptom. Holder, cable, etc., feel hottest; abnormal resistance found in these parts by "drop" method.

Remedy. Improve the connections.

Cause 6. Arcing or short circuit in commutator. This may occur across mica or insulation between bars or other parts.

Symptom. Burnt spot between parts; spark appears in the insulation when current is put on.

Remedy. Pick out the charred particles; take commutator apart and repair; or put on new commutator.

Cause 7. Carbon brushes heated by current.

Symptom. Brushes hotter than other parts.

Remedy. Use carbon of higher conductivity. Let the brush-holder grip brush closer to commutator, so as to reduce the length of brush through which the current must pass. Use larger brushes or a greater number.

III. Heating of Armature

Cause 1. Excessive current in armature coils.

Symptom and Remedy. Symptom and Remedy the same as in case of I, Cause 1.

Cause 2. Short-circuited armature coils.

Symptom and Remedy. Symptom and Remedy the same as in case of I, Cause 5. See also Cause 7.

Cause 3. Moisture in armature coils.

Symptom. The detection of moisture in armature coils is a difficult matter without measuring the insulation resistance. Sometimes a careful inspection of the armature may reveal the presence of moisture, but on large machines the safest method is to apply an insulation test.

Remedy. When the insulation test reveals the presence of moisture, the armature, if small, may be baked in an oven sufficiently warm to expel all moisture, but not hot enough to injure the insulation. If the armature is large it can not readily be removed and placed in an oven. The machine may then be connected to run as a generator or motor, at reduced voltage, in order not to break down the insulation which is weakened by the moisture present, the voltage being gradually brought up as the insulation increases. If more convenient, a current controlled by resistance or otherwise at, say, half of rated value may be sent through a moist armature or field winding to dry it out.

Cause 4. Foucault currents in armature core.

Symptom. Iron of armature core hotter than coils after a short run, and considerable power required to run armature when field is magnetized and there is no load on armature. This can be distinguished from Cause 2 by absence of sparking and absence of excessive heat in a particular coil or coils after a short run. (See "Stray Power Tests".)

Remedy. Armature core should be laminated more perfectly, which is a matter of first construction.

Cause 5. One or more reversed coils on one side of armature. This will cause a local current to circulate around armature.

Symptom. Excessive current when running free, but no particular coil heated more than others. If a moderate current is applied to each coil in succession by touching wires carrying current to each pair of adjacent commutator-bars, a compass needle held over the coils will behave differently when the reversed coil is reached. In a motor, the half of armature containing the reversed coils is heated more than the others.

NOTE.—Any excess of current taken by an armature when running *free* as a motor, whatever the cause, must be converted into heat by some defect in the machine; hence the "free current" is the simplest and most complete test of efficiency and perfect condition.

Remedy. Reconnect the coil to agree with the others.

Cause 6. Heat conveyed from other parts.

Symptom. Other parts hotter than armature. Start with machine cool, and see if other parts heat first.

Remedy. See II, IV, and V.

Cause 7. Flying cross in armature conductor.

Symptom and Remedy. Symptom and Remedy similar to the case of sparking (I, Cause 11), except that reference here is to the insulation of the conductors.

IV. Heating of Field Magnets

Cause 1. Excessive current in field circuit.

Symptom. Field coils too hot to keep the hand on. Their temperature more than 50 degrees centigrade above that of room by resistance test or by thermometer.

Remedy. In the case of a shunt-wound machine, decrease the voltage at terminals of field coils; or increase the resistance in field circuit by winding on more wire or putting resistance in series. In the case of a series-wound machine, shunt a portion of, or otherwise decrease, the current passing through field; or take a layer or more of wire off the field coils; or rewind with coarser wire. This trouble might be due to a short circuit in field coils in the case of a shunt-wound dynamo or motor, and would be indicated by the pole piece with the short-circuited coil being weaker than the others. This coil is cooler than the others; in fact, if completely short-circuited, it is not heated at all. This condition can be remedied only by rewinding the short-circuited coil. Measure resistance of the field coils to see if they are nearly equal. (See "Drop Method", p. 88.) If the difference is considerable (say, more than 5 or 10 per cent) it is almost a sure sign that one coil is short-circuited or double-grounded.

Cause 2. Foucault currents in pole pieces or field cores.

Symptom. The pole pieces hotter than the coils after a short run. When making the comparison, it is necessary to keep the hand on the coils some time before the full effect is reached, because the coils are insulated and the pole pieces are bare metal, and even then the coils will not feel so hot, although their actual temperature may be higher, if measured by a thermometer.

Remedy. This trouble is due to faulty design of toothed-armature machines, which can be corrected only by rebuilding, or is caused by fluctuations in the current. The latter can be detected, if the variations are not too rapid, by putting an ammeter in circuit; or rapid variations may be felt by holding a piece of iron near the pole pieces and noting whether it vibrates.

Cause 3. Moisture in field coils.

Symptoms. The field circuit tests lower in resistance than normal in that type of machine; and in the case of shunt-wound machines the field takes more than the ordinary current. Field coils steam when hot, or feel moist to the hand. The insulation resistance also tests low.

Remedy. Remove and bake field coils as in the case of small armatures. On large machines the field coils may be connected in parallel-series groups so as to get approximately one-half normal current at reduced voltage. As the insulation resistance rises, the current may be brought up to normal value. (See "Heating of Armature", III, 3.)

V. Heating of Bearings

The cause should be found and removed promptly, but heating of the bearings can be reduced temporarily by applying cold water or ice to them. This is allowable only when absolutely necessary to keep running; and great care should be taken not to allow any water to get upon the commutator, armature, or field-coils, as it might short-circuit or ground them. If the bearing is very hot, the shaft should be kept revolving slowly while the bearing is flushed with fresh oil, as it might "freeze". or stick fast, if stopped entirely.

Cause 1. Lack of oil.

Symptom. Oil reservoir empty. Self-oiling rings fail to turn with shaft. Shaft and bearing look dry, and often there is an odor of hot oil.

Remedy. Loosen bearing cap screws and then retighten them with fingers only. This will give the bearings a little clearance. Open the drain cock on the side of the bearing and pour fresh oil in top of bearing, allowing the oil to run through the bearing until temperature is reduced to a safe value.

Cause 2. Grit or other foreign matter in bearings.

Symptom. Best detected by removing shaft or bearings and examining both. Any grit can, of course, be felt easily, and will also cut the shaft.

Remedy. Remove shaft or bearing, clean both very carefully, and see that no grit can get in. The oil should be perfectly clean; if it is not, it should be filtered.

Cause 3. Shaft rough or cut.

Symptom. Shaft will show grooves or roughness and will probably revolve stiffly.

Remedy. Turn shaft in lathe; or smooth with fine file; and see that bearing is smooth and fits shaft.

Cause 4. Bearings fit too tight.

Symptom. Shaft hard to revolve. Hot spots develop in bearing.

Remedy. Remove the bearing and scrape the spots that show wear; then coat the shaft with a thin layer of red lead and place the bearing on the shaft. Move the bearing back and forth and when it is removed the high spots will be covered with the red lead. These should be scraped off, and the process repeated until a snug-fitting bearing is obtained.

Cause 5. Shaft "sprung" or bent.

Symptom. Shaft hard to revolve, and usually sticks much more in one part of revolution than in another.

Remedy. It is very difficult to straighten a bent shaft. It might be bent back or turned true, but probably a new shaft will be necessary.

Cause 6. Bearings out of line.

Symptom. Shaft hard to revolve, but is much relieved by slightly loosening the screws that hold the bearings in place, when machine is not running and when belt, if any, is taken off.

Remedy. Loosen the bearings by partly unscrewing bolts or screws holding them in place, and find their easy and true position, which may require one of them to be moved either sideways or up or down; then ream the holes of that bearing, or raise or lower it, as may be necessary, to make it occupy the right position when the screws are tightened. The armature, however, must be kept in the center of the space between the pole pieces, so that the clearance is uniform all around. (See Cause 9.)

Cause 7. Thrust or pressure of pulley, collar, or shoulder on shaft against one or both of the bearings.

Symptom. Move shaft back and forth with a stick applied to the end while revolving, and note if the collar or shoulder tends to be pushed or drawn against either bearing. It is usually desirable that a shaft should move freely back and forth about an eighth of an inch, to make commutator and bearings wear smoothly.

Remedy. Line up the belt; shift collar or pulley; turn off shoulder on shaft, or file off bearing, until the shoulder does not touch when running, or until pressure is relieved.

Cause 8. Too great a load or strain on the belt.

Symptom. Great tension on belt. In this case the pulley bearing will probably be very much hotter than the other, and also worn elliptical, as indicated in Fig. 93, in which case the shaft can be shaken in the bearing in the direction of the belt pull, when the belt is off, provided the machine has been running long enough to wear the bearings.

Fig. 93. Diagram Showing Elliptical Bearing Due to Wear

Remedy. Reduce load or tension, or use larger pulleys and lighter belt, so as to relieve side strain on shaft. (See "Belting", p. 7.)

Cause 9. Armature off of center, producing much greater magnetic pull toward nearer side.

Symptom. Heating of bearing on side where air gap is smallest. Poor commutation due to unequal air gap.

Remedy. The fault is due either to some inherent defect in the machine, to the faces of the poles not being concentric with the armature, to faulty erection of the machine, or to wear on the bearings.

In most cases this trouble can be remedied by raising or lowering either the bearing pedestals or the magnet frame feet, by means of their sheet-iron shims, to obtain the proper air gap vertically. To obtain the correct air gap in a horizontal direction, there is usually enough play in the foundation bolt holes in large machines. On small machines the proper adjustment is usually made by the manufacturer. If the bearings have worn very much it is, of course, advisable to re-babbitt them to restore the correct air gap between armature and fields.

VI. NOISY OPERATION

Cause 1. Vibration due to armature or pulley being out of balance.

Symptom. Strong vibration felt when the hand is placed upon the machine while it is running. Vibration changes greatly if speed is changed, and sometimes almost disappears at certain speeds.

Remedy. Armature or pulley must be perfectly balanced by securely attaching lead or other weights on the light side, or by drilling or filing away some of the metal on the heavy side. The easiest method of finding in which direction the armature is out of balance is to take it out, and to rest the shaft on two parallel and horizontal A-shaped metallic tracks sufficiently far apart to allow the armature to go between them, Fig. 94. If the armature is then slowly rolled back and forth, the heavy side will tend to turn downward. The armature and pulley should always be balanced separately. An excess of weight on one side of the pulley and an equal excess of weight on the opposite side of the armature will not produce a balance while running, though it does when standing still; on the contrary, it will give the shaft a strong tendency to "wobble". A perfect balance is obtained only when the weights are directly opposite, i.e., in the same line perpendicular to the shaft.

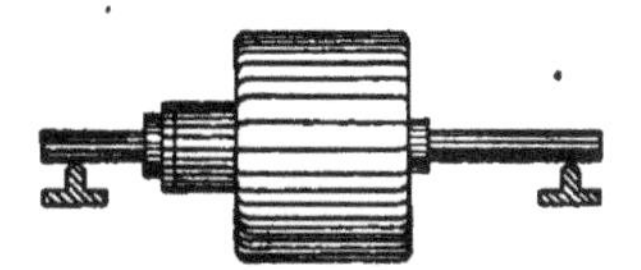

Fig. 94. Method of Putting Armature in Proper Balance

Cause 2. Armature strikes or rubs against pole pieces.

Symptom. Easily detected by placing the ear near the pole pieces; or by examining armature to see if its surface is abraded at any point; or by examining each part of the space between armature and field as armature is slowly revolved, to see if any portion of it touches, or is so close as to be likely to touch, when the machine is running. In small machines, the armature may be turned by hand to find out whether it sticks at any point.

Remedy. See V, 9.

Cause 3. Shaft collar or shoulder, hub or edge of pulley, or belt, strikes or scrapes against bearings.

Symptom. Rattling noise, which stops when the shaft or pulley is pushed lengthwise away from one or the other of the bearings. (See V, Cause 7.)

Remedy. Shift the collar or pulley, turn off the shoulder on the shaft, file or turn off the bearing, move the pulley on the shaft, or straighten the belt, until there is no more striking, and the noise ceases.

Cause 4. Rattling due to looseness of screws or other parts.

Symptom. Close examination of the bearings, shaft, pulley, screws, nuts, terminals, etc., or touching the machine while running, or shaking its parts while standing still, shows that some parts are loose.

Remedy. Tighten up the loose parts, and be careful to keep them all properly set up. It is easy to guard against the occurrence of this trouble, which is very common, by simply examining the various screws and other parts each day before the machine is started. Electrical machinery being usually high speed, the parts are particularly liable to shake loose. A worn or poorly fitted bearing might allow the shaft to rattle and make a noise, in which case the bearing should be refitted or renewed.

Cause 5. Singing or hissing of brushes. This is usually occasioned by rough or sticky commutator (see Trouble I, Causes 3 and 10), or by brushes not being smooth, or by the layers of a copper brush not being held together in place. With carbon brushes, hissing will be caused by the use of carbon that is gritty or too hard. Vertical carbon brushes, or brushes inclined against the direction of rotation, are liable to squeak or sing. Occasionally, a new machine will make a noise that is reduced after the machine has been run for some time.

Symptom. Sounds of high pitch are easily located by placing the ear near the commutator while it is running, and by lifting off the brushes one at a time—provided there are two or more brushes in each set, so that the circuit is not opened by lifting a single brush. If there is no current there is, of course, no objection to raising the brushes.

Remedy. Apply a *very little* oil to the commutator with a rag on the end of a stick. Adjust the brushes or smooth the commutator by turning, or by using fine sandpaper, being careful to clean thoroughly afterwards. Carbon brushes are liable to squeak in starting up or at low speed. This squeaking decreases at full speed, and can generally be stopped altogether by moistening the brushes with oil, care being taken not to have any excess of oil. Running the machine without load for some time usually reduces this trouble.

Cause 6. Flapping or pounding of belt joint or lacing against pulley. (Fig. 95.)

Fig. 95. Typical Bad Belt Joints

Symptom. Sound repeated twice for each complete revolution of the belt, which is much less frequent than any other generator or motor sound, and can easily be detected or counted.

Remedy. Endless belt or smoother joint. (See "Belting", p. 8.)

Cause 7. Slipping of belt on pulley due to overload.

Symptom. Intermittent squeaking noise.

Remedy. Tighten the belt or reduce the load. A wider belt or larger pulley may be required. Powdered rosin may be put on the belt to increase its adhesion; but such a procedure is a makeshift, is injurious to the belt, and should be adopted only when necessary. (See "Belting", p. 7.)

Cause 8. Humming of armature-core teeth as they pass the pole pieces.

Symptom. Pure humming sound less metallic than Cause 5.

Remedy. Slope or chamfer the ends of the pole pieces so that each armature tooth does not pass the edge of the pole piece all at once. Decrease the magnetization of the fields. Increase the air gap or reduce the distance between the teeth. But these are nearly all matters of first construction and are made right by good manufacturers.

Cause 9. Humming due to alternating or pulsating current.

Symptom. This gives a sound similar to that in the preceding case. The two can be distinguished, if necessary, by determining whether the note given out corresponds to the number of alternations or to the number of armature teeth passing per second. Usually the latter is considerably greater than the former.

Remedy. This trouble is confined to alternating apparatus, and its effects can be reduced by proper design and by mounting the machine so as to deaden the sound as far as possible.

NOTE.—It often happens that a generator or motor seems to make a noise, which in reality is caused by the engine or other machine with which it is connected. Careful listening with the ear close to the different parts will show exactly where the noise originates. A very sensitive method of locating a noise or vibration is to place one end of a short stick between the teeth, and press the other end squarely against the various parts, to ascertain which particular one gives the greatest vibration.

VII. SPEED NOT RIGHT

This is generally a serious matter in either a generator or a motor, and it is always desirable and often imperative to shut down immediately, and make a careful investigation.

Speed Too Low

Cause 1. Overload. (See I, Cause 1.)

Symptom. Armature runs more slowly than usual. Bad sparking at commutator. Ammeter indicates excessive current. Armature heats. Belt very tight on tension side.

Remedy. Reduce the load on machine, increase the diameter of driving pulley, or decrease the diameter of driven pulley. If necessary to relieve strain of overload, temporarily decrease the voltage on either a generator or a motor.

Cause 2. Short circuit or ground in armature.

Symptom and Remedy. Symptom and remedy the same as in case of III, Cause 2 and Cause 6.

Cause 3. Armature strikes pole pieces.

Symptom and Remedy. Symptom and remedy the same as in the case of VI, Cause 2.

Cause 4. Shaft does not revolve freely in the bearings.

Symptom and Remedy. Symptom and Remedy the same as for V, all cases.

Cause 5. Poor contact in energy circuit. (This applies to single-phase repulsion motors only.)

Symptom. As explained previously, the repulsion motor has one set of brushes short-circuited. These are the energy brushes and if there is a poor contact in this circuit, it will cause a reduction in the output and in the speed of the motor. If this circuit is opened entirely, the motor will stop.

Remedy. Test out this circuit by any resistance method, find the faulty connection, and repair it.

Speed Too High or Too Low

Cause 6. Field magnetism weak. This has the effect, on a constant-voltage circuit, of making a motor run too fast if lightly loaded, or too slow if heavily loaded. It makes a generator fail to "build up" or excite its field, or give the proper voltage in any case.

Symptom and Remedy. Symptom and remedy the same as in the case of "Sparking", Cause 8. (See VII, Cause 7; also IX.)

Cause 7. Too high or too low voltage on the circuit.

Symptom. This would cause a motor to run too fast or too slow, respectively. It can be shown by measuring the voltage of the circuit.

Remedy. The central station or generating plant should be notified that voltage is not right.

Speed Too High

Cause 8. Motor too lightly loaded.

Symptom. A series-wound motor on a constant-potential circuit runs too fast, and may speed up to the bursting point if the load is very much reduced or removed entirely (by the breaking of the belt, for example).

Remedy. Care should be exercised in using a series motor on a constant-potential circuit, except where the load is a fan, pump, or other machine that is *positively* connected or geared to the motor so that there is no danger of its being taken off. A shunt- or compound-wound motor should be used, if the load is likely to be thrown off.

Cause 9. Poor contact or open circuit in compensating circuit. (This applies to single-phase repulsion motors only.)

Symptom. No-load speed may be one and one-half times rated speed.

Remedy. Look for poor contact or open circuit and test by resistance method. Repair contact or break.

VIII. MOTOR STOPS OR FAILS TO START

This is an extreme case of the previous class ("Speed Not Right"), but is separated because it is more definite and permits of quicker diagnosis and treatment. This heading does not, of course, apply to generators, since any trouble in setting these in motion is usually outside of the machine itself.

Cause 1. Great overload. A slight overload causes motor to run slowly, but an extreme overload will, of course, stop it entirely or "stall" it. (See I, Cause 1.)

Symptom. Of course the chief symptom is the stopping of the motor. On a constant-potential circuit the current is excessive, and safety fuse blows or circuit breaker opens. In their absence or failure, armature is burnt out.

Remedy. Turn off current instantly, reduce or take off the load, replace the fuse or circuit breaker, if necessary, and turn on current again just long enough to see if trouble still exists; if so, take off more load.

Cause 1a. Load too great (a. c. motors). An induction motor if overloaded will slow down somewhat, and finally stop altogether. It may also fail to start for the same reason. A synchronous motor will maintain its speed with any load it can carry. It will not start under too much load or will pull out of step and stop if overloaded while running.

Symptom. The symptoms may be the same as in Cause 1.

Also there will be a pronounced humming indicating excessive current in the windings.

Remedy. Apply remedies as given in Cause 1.

Cause 2. Very excessive friction due to shaft, bearings, or other parts being jammed, or armature touching pole pieces.

Symptom. Similar to previous case, but distinguished from it by the fact that the armature is hard to turn even when load is taken off. Examination shows that the shaft is too large or is bent or rough, that the bearing is too tight, that the armature touches pole pieces, or that there is some other impediment to free rotation. (See V and VI.)

Remedy. Turn current off instantly, ascertain and remove the cause of friction, turn on the current again just long enough to see if trouble still exists; if so, investigate further.

Cause 3. Circuit open. This may be due to (a) safety-fuse blown or circuit breaker open; (b) wire in motor broken or slipped out of connections; (c) brushes not in contact with commutator; (d) switch open; (e) circuit supplying motor open; (f) failure at generating plant.

Symptom. Distinguished from Causes 1 and 2 by the fact that if the load is taken off, the motor still refuses to start, and yet armature turns freely.

On a constant-potential d. c. circuit, the field circuit alone of a shunt motor may be open, in which case the pole pieces are not strongly magnetic when tested with a piece of iron, and there is a dangerously heavy current in the armature; if the armature circuit is at fault, there is no spark when the brushes are lifted; and if both are without current, there is no spark when switch is opened. One should be very careful if there is no field magnetism or even if it is very weak, as a motor is liable to be burnt out if the current is then thrown upon the armature.

Remedy. Turn off current instantly. Examine safety fuse, circuit breaker, wires, brushes, switch, and circuit generally, for break or fault. If none can be found, turn on switch again for a moment, as the trouble may have been due to a temporary stoppage of the current at the station or on the line. If motor still seems dead, test separately armature, field coils, and other parts of circuit for continuity with a magneto, or a cell of battery and an electric bell, to see if there is any break in the circuit. (See "Instructions for Testing", pages 88 to 93.

One of the simplest ways to find whether the circuit has current in it and to locate any break, is to test through an incandescent lamp. Two and five lamps in series should be used on 220- and 500-volt circuits, respectively.

Cause 3a. Circuit open. (a. c. motors.)

Symptom. In an a. c. motor there may be a pronounced humming, showing that there is current in the motor, and yet it will not start. Even with no load the motor will not start, although the rotor may be turned by hand.

Remedy. Turn off current. Check connections external to motor to locate any open circuit. If one phase only is open, allowing only single-phase current to reach the motor, it will not start and the fault can be repaired and the motor started. If no open circuit can be found, try again to start the motor, since the circuit from the power house may have been temporarily interrupted. If the motor still fails to start, the trouble must be from an open circuit within the motor and the usual tests for this condition will have to be made. It may be necessary in this case to remove some coils and replace them with new ones. If it is an induction motor with an external resistance in the rotor circuit, a test should be made for grounds or short circuits in the resistance. Such grounds or shorts can usually be easily found and removed.

In addition to the above, there may be an open circuit in some auxiliary apparatus, such as a transformer or starting compensator. Tests for open circuits should be applied to these auxiliaries as well as to the motor and the breaks or loose connections repaired.

Cause 4. Wrong connection or complete short circuit of field, armature, switch, etc.

General Symptom. Distinguished from Causes 1 and 2 in the same way as Cause 3, and differs from Cause 3 in the evidence of strong current in motor.

Symptom on a constant-potential circuit. If current is very great it indicates a short circuit. If the field is at fault it will not be strongly magnetic.

The possible complications of wrong connections are so great that no exact rules can be given. Carefully examine and make sure of the correctness of all connections (see "Diagram of Connections" accompanying the machine). This trouble is usually inexcusable, since only a competent person should ever set up a machine or change its connections.

Symptoms in the three-wire (220-volt direct-current) system. Several peculiar conditions may exist, as follows:

(a) The generator or generators on one side of the system may become reversed, so that both of the outside wires are positive or negative. In that case a motor fed in the usual way from the two outside wires will get no current, but lamps connected between the neutral wire and either of the outside wires will burn as usual.

(b) If one of the outside wires is open by the blowing of a fuse, an accidental break, or other cause, then a motor (220-volt) beyond the break can get some current at 110 volts through any lamps that may be on the same side of the break as itself, and on the same side of the system as the conductor that is open. These lamps will light up when the motor is connected, but the motor will have little or no power unless the number of lamps is large.

(c) If the neutral or middle wire is open, a motor connected with the outside wires will run as usual; but lamps on one side of the system will burn more brightly than those on the other side, unless the two sides are perfectly balanced.

(d) If one of the outside wires becomes accidentally grounded, a 110-volt generator, motor, or other apparatus, also grounded and connected to the other outside wire, will receive 220 volts, which will probably burn it out.

Polyphase Induction Motor Operates Single-Phase

Cause. Open circuit in one lead.

Symptom. Motor will not carry full rated load and heats too much. Motors are often started with the fuses cut out. In that case the motor would start satisfactorily and a blown fuse in one leg would not be discovered. After once up to speed, the motor could be thrown over to running position and would run single-phase, but would carry only a part of its load (about 70 per cent for a three-phase motor).

Remedy. Locate the break in the circuit and replace the fuse or repair the break.

Synchronous Motor Fails to Reach Synchronous Speed

Cause 1. Short-circuited field.

Symptom. Motor starts, but comes up to a speed far below synchronism and sticks at that speed. This speed is not high enough to allow the starting switch to be thrown over into full voltage position.

Remedy. Open the circuit. Examine the revolving field for external short circuits, or shorts in the connections between spools and collector rings. If a resistance is used across the fields in starting, it may be short-circuited, thus shorting the fields. Test the field circuit for grounds which might cause short circuits. Remove any shorts discovered. The motor will then come within about 5 per cent of full speed on the starting tap of the transformer or compensator and can be thrown on full voltage and given current in the fields.

Cause 2. Too much load.

Symptom. The motor may fail entirely to start, or may come up to sub-synchronous speed as in the previous case.

Remedy. Remove some of the load, or arrange a clutch so that the load can be applied after the motor is on the line and running at full speed. If neither of these methods can be used, an auxiliary induction motor may be belted to the motor to help it up to speed.

Synchronous Motor Flashes Across Collector Rings or End Field Coils at Starting

Cause. High voltage induced in fields.

Symptom. When starting switch is first thrown in, a bad flash appears at the collector rings or field coils.

Remedy. Insert a properly designed resistance so that it is across the fields during the starting operation. This resistance will prevent the rise in voltage, which, in some machines, may reach a value as high as 2500 or 3000 volts.

IX. DYNAMO FAILS TO GENERATE

This trouble is almost always caused by the inability of a generator to "build up" or excite its field-magnetism sufficiently. The proper starting of a self-exciting machine requires a certain amount of residual magnetism, which must be increased to full strength by the current generated in the machine itself. This trouble is not likely to occur on a separately-excited machine; and if it does, it is usually due to the exciter failing to generate, and therefore amounts to the same thing.

Cause 1. Residual magnetism too weak or destroyed.

This may be due to (a) vibration or jar; (b) proximity of another generator; (c) earth's magnetism; (d) accidental reversed current through fields, not enough to completely reverse magnetism. The complete reversal of the residual magnetism in any machine will not prevent its generating, but will only make it build up of opposite polarity. Sometimes reversal of residual magnetism may be very objectionable, as in case of charging storage batteries, but, although the popular supposition is to the contrary, it will not cause the machine to fail to generate.

Symptom. Little or no magnetic attraction when the pole pieces are tested with a piece of iron.

Remedy. Send a magnetizing current from another machine or battery through the field coils, then start and try the machine; if this fails, apply the current in the opposite direction, since the magnets may have enough polarity to prevent the battery building them up in the direction first tried.

Shift the brushes backward in a generator or forward in a motor to make armature magnetism assist field.

Cause 2. Reversed connections or reverse direction of rotation.

Symptom. When running, pole pieces show no attraction for a piece of iron. The application of external current cannot be made to start the machine, as in the case of Cause 1, because, whichever way the field may be magnetized, the resulting current generated by armature opposes and destroys the magnetism.

Remedy. (a) Reverse either armature connections or field connections, *but not both.* (b) Move brushes through 180 degrees for two-pole, 90 degrees for four-pole machines, etc. (c) Reverse direction of rotation. After each of the above are tried, the field may have to be built up with a battery or other current, since the causes in this case tend to destroy whatever residual magnetism may have been present.

Cause 3. Short circuit in the machine or external circuit. This applies to a shunt-wound machine, and has the effect of preventing the voltage and the field magnetism from building up.

Symptom. Magnetism weak, but still quite perceptible.

Remedy. If the short circuit is in the external circuit, opening the latter will allow the dynamo to build up and generate full voltage. If the short circuit is within the machine, it should be found by careful inspection or testing. In either of these cases, do not connect the external circuit until short circuit is found and eliminated. A slight short circuit, such as that caused by copper dust on the brush holder or commutator, may prevent the magnetism of a shunt machine from building up. (See "Sparking", I, Causes 5 and 8.) Too much load might prevent a shunt dynamo from building up its field magnetism, in which case the load, or some of it, should be disconnected in starting.

Cause 4. Field coils opposed to each other.

Symptom. Upon passing a current from another source of supply, the following symptom will exist: If the pole pieces of a bipolar machine are approached with a compass or other freely suspended magnet, they both attract the same end of the magnet, showing them to be of the same polarity, whereas they should always be of opposite polarity.

For similar reasons the pole pieces are magnetic when tested separately with a piece of iron, but show less attraction when the same piece of iron is applied to both at once, in which latter case the attraction should be stronger. In multipolar machines these tests should be applied to consecutive pole pieces.

Remedy. Reverse the connections of the incorrectly connected coils in order to make the polarity of the pole pieces opposite. The pole pieces should be alternately north and south.

Cause 5. Open circuit. This may be due to (a) broken wire or faulty connection in machine; (b) brushes not in contact with commutator; (c) safety fuse melted or absent; (d) switch open; (e) external circuit open.

Symptom. If the trouble is merely due to the switch or external circuit being open, the magnetism of a shunt generator may be at full strength, and the machine itself may be working perfectly; but if the trouble is in the machine, the field magnetism will probably be very weak.

Remedy. Make a very careful examination for open circuit; if not found, test separately the field coils, armature, etc., for continuity, with magneto or cell of battery and electric bell. (See "Instructions for Testing", p. 92, also VIII, Cause 3.)

A break, poor contact, or excessive resistance in the field circuit or field rheostat of a shunt dynamo will also make the magnetism weak and prevent its building up. This may be detected and overcome by cutting out the rheostat for a moment by connecting the two terminals of the field coils to the two brushes respectively, care being taken not to make a short circuit.

A break or abnormally high resistance anywhere in the external circuit of a series-wound dynamo will prevent it from generating, since the field coil is in the main circuit. This may be detected and overcome by short-circuiting the machine for a moment in order to start up the magnetism.

Either of these two remedies by short-circuiting should be applied very carefully, and not until the pole-pieces have been tested with a piece of iron to make sure that the magnetism is weak.

Cause 6. Brushes not in proper position.

Symptom. The strength of the field increases, and the voltage of the generator builds up.

Remedy. It sometimes happens that brushes are not set at the proper point on commutator and, if the residual magnetism is already weak, the machine may fail to build up. Almost all modern generators made in this country have the brushes set about opposite the center of the pole piece. Generators require the brushes to be shifted in the direction of rotation of the armature; motors require a backward shift to the brushes. Occasionally a generator may be found with armature wound in such a way that the proper position of the brushes is midway between the adjacent pole tips.

X. VOLTAGE OF GENERATOR NOT RIGHT

Cause 1. Speed too low. (See VII.)

Remedy. Increase speed of the prime mover if possible; when this cannot be done, decrease the diameter of the driven pulley or increase the diameter of the driving pulley, preferably the latter.

Cause 2. Field magnetism weak.

Symptom and Remedy. See I, Cause 8.

Cause 3. Brushes not in proper position.

Symptom and Remedy. See I, Cause 2.

Cause 4. Machine overloaded.

Symptom and Remedy. See I, Cause 1, and VII, Cause 1; also increase field excitation, if possible.

Cause 5. Short-circuited armature coil or coils.

Symptom and Remedy. See I, Cause 5.

Cause 6. Reversed armature coil or coils.

Symptom and Remedy. See I, Cause 5.

Voltage Too High

Cause 7. Speed too high.

Remedy. Apply the reverse of treatment given in Cause 1.

Cause 8. Field magnetism too powerful.

Remedy. Increase resistance of shunt-field circuit, by means of a shunt-field rheostat.

Cause 9. Machine compounds too much.

Remedy. Decrease resistance of series-field shunt. (See "Compound-Wound Dynamos", page 17, Part I.)

In commutating-pole generators it sometimes happens that the commutating poles themselves exert a compounding effect; especially is this true if the brushes are shifted slightly back of the proper neutral point.

TYPES OF GENERATORS AND MOTORS

PART I

INTRODUCTION

Dynamo-electric machinery is the generic name given to the class of machines that employ relative motion between an electric conductor and a magnetic field for the purpose of converting mechanical power into electrical power, or the reverse.

Generators and motors constitute the two largest and most important subdivisions of this class of machinery. A generator transforms mechanical power into electrical power, while a motor transforms electrical power into mechanical power. Originally the name dynamo-electric machine was used to designate an electric generator. On account of its length, this term was soon shortened to that of dynamo. Recently the name generator has supplanted that of dynamo, it being considered preferable on account of its rather self-evident meaning, while the term dynamo-electric machine is now used in the extended meaning given above.

METHODS OF DRIVING

GENERATORS

The methods of driving generators in this country today are: direct connection to the prime mover; by belting (including rope-driving, which is rather exceptional), and by gearing.

Direct-Connected Generators. Direct-connected or direct-driven generators are manufactured in combination with steam engines, gas engines, water wheels, and steam turbines; the generator being, in any particular case, of the direct-current or the alternating-current type as desired.

Steam- or Gas-Engine Driven. A direct steam- or gas-engine driven generator may be arranged in one of three different ways. First, it may have its revolving part carried by the engine shaft proper and is usually then denoted as an engine-type generator. Such a generator, as shown in Fig. 1, has no shaft and no bearings

Fig. 1. Allis-Chalmers Type I Generator and Corliss Engine
Courtesy of Allis-Chalmers Manufacturing Company

Fig. 2. Allis-Chalmers Generator Direct-Connected to "Ideal" Engines
Courtesy of Allis-Chalmers Manufacturing Company

of its own, both shaft and bearings being provided by the engine builder. Second, the engine shaft may be extended out to one side of the engine, which extension then carries the revolving part of the generator, as shown in Fig. 2. There is then provided in this arrangement an extra bearing, usually called an *outboard bearing*. Third, both engine and generator may be complete in themselves and their shafts directly coupled together by a suitable coupling either rigid or flexible. This arrangement is illustrated in Figs. 3 and 4. In all

Fig. 3. Giles Engine Direct-Connected to Crocker-Wheeler Generator
Courtesy of Crocker-Wheeler Company

cases a flywheel is provided, increasing the inertia of the rotating parts and thus increasing the uniformity of rotation.

Water-Wheel Driven. A direct water-wheel driven generator may be arranged with a vertical or with a horizontal shaft. Those with a vertical shaft are usually of the alternating-current type, ranging in sizes from about 100 up to 10,000 kilowatts in capacity, as shown in Fig. 5, although they are sometimes direct-current generators, as shown in Fig. 6. When the revolving part of the generator is mounted directly upon the vertical shaft of a turbine, some form of thrust or step bearing becomes necessary to support the combined weight of the revolving generator and turbine parts. The pressure on the bearings is decreased in some instances by balancing all or a part of the weight, either by the upward pressure of the water in the turbine or by the magnetic pull between the field

Fig. 4. Crocker-Wheeler Generator Driven by Producer Gas Engine
Courtesy of Crocker-Wheeler Company

and the armature of the generator. Fig. 7 shows the arrangement of horizontal shaft turbines directly connected to a generator, and Fig. 8 shows a direct-current generator driven by a turbine of the impulse type. Direct-current generators direct-driven by water wheels are in most cases railway generators.

Steam-Turbine Driven. Direct steam-turbine driven generators are similarly manufactured with either vertical or horizontal shafts. An example of the vertical type is the Curtis steam turbine unit manufactured by the General Electric Company and shown in Fig. 9. The generator of this type of unit is, however, as manufactured at present, always a three-phase alternator. Illustrations of horizontal-shaft steam-turbine units are given in Figs. 10 and 11.

Fig. 5. Bank of Water-Wheel Driven Alternators
Courtesy of General Electric Company

Belted Generators. Generators that are not direct-connected to their prime movers are almost always driven by belt, or in rather special cases by some form of rope drive or silent chain drive. Belt-driven generators, up to 150-kw. capacity, usually carry a metal, wood, or paper pulley that merely overhangs the main bearings. If the generator is larger, or from 150- to 500-kw. capacity, a third bearing is provided to relieve the bending strain occasioned by the belt pull. The prime movers used in connection with belted generators are generally only steam engines, gas engines, and water wheels. In the smaller sizes in isolated plants, however, various internal

Fig. 6. Generators Connected Direct on Top of Turbine Shaft
Courtesy of Trump Manufacturing Company

combustion engines are found driving belted generators. The steam turbine is never belted to a generator, since in the smaller powers it rotates at very high speed, rendering the use of belting impracticable.

Fig. 7. Two 2400-Volt Generators Direct-Connected to Water Wheels
Courtesy of Crocker-Wheeler Company

Geared Generators. Gearing is only used for driving generators in the following cases: First, vertical-shaft water wheels and horizontal-shaft generators are often connected, in case of low heads, by means of bevel gearing, the generator shaft revolving at a higher speed than the turbine. Second, in smaller sizes, steam turbines

Fig. 8. Doble Water Turbine Direct-Connected to Generator

Fig. 9. Curtis Steam Turbine Unit
Courtesy of General Electric Company

and generators are connected by means of gearing, the generator shaft revolving at a lower speed than the steam-turbine shaft. An example of this method of connection is given in Fig. 12. Third,

Fig. 10. Kerr Steam Turbine Generating Unit
Courtesy of Kerr Turbine Company

direct-current generators used as exciters for steam-turbine driven alternators are sometimes geared to the shaft of the unit.

Selection of Type of Generator Drive. At the present time, the best modern practice employs direct-driven generators whenever

Fig. 11. Terry Steam Turbine Direct-Connected to Generator
Courtesy of Terry Steam Turbine Company

possible, and only the smaller units are ever belt-driven, since even the best system of belting is unsatisfactory for transmitting large amounts of power. Direct driving necessitates, in general, a larger, heavier, and, therefore, more expensive generator for the same

Fig. 12. De Laval Double-Geared Turbine Driving Two D. C. Generators
Courtesy of De Laval Steam Turbine Company

capacity; but the compactness, simplicity, positive action, and general advantages of direct driving are so great that in most cases they warrant the increased cost. An important advantage possessed by the direct-connected generator is the small floor space required. In some instances this advantage allows a saving in the cost of the building and in the real estate, and more than counterbalances the increased cost of the generating unit. In belt driving, the distance between the centers of the engine and generator shaft should be at least three times the diameter of the engine pulley to insure satis-

Fig. 13. Two Induction Motors Direct-Connected to Centrifugal Pumps
Courtesy of General Electric Company

factory working of the belt. In some cases, however, the belt-driven generator may be preferable, particularly in the smaller sizes, where only the cost of the units is to be considered, or on account of the ease with which the generator can be adjusted to an engine already in use.

MOTORS

Similarity to Generators. The same methods as are employed for driving generators are also used in connecting motors to the work they perform. Direct connection is resorted to wherever feasible. As a result, many direct-current constant-speed motors are identical in appearance, as regards their general outlines, with direct-driven generators. In fact, manufacturers actually employ the same frames in many cases for direct-current generators and motors, merely changing the windings to produce the required char-

Fig. 14. Example of Rope Drive
Courtesy of Crocker-Wheeler Company

acteristics. Alternating-current generators and synchronous motors are really identical, that is, the same machine may be used for either purpose as desired. Two direct-connected motors are shown in Fig. 13.

Belted motors are restricted to the smaller sizes, due to the limitations of belt transmission. The employment of rope drive, silent chain drive, and gearing is much more common than in the case of generators. Illustrations of the last three methods are given in Figs. 14, 15, and 16.

Fig. 15. Motors Driving Jack Shafts Through Silent Chain
Courtesy of General Electric Company

Fig. 16. Induction Motor Geared to Belt Conveyor
Courtesy of General Electric Company

DIRECT-CURRENT TYPES

General Classification. The various types of dynamo-electric machinery fall naturally under the two main divisions of alternating-current and direct-current machines. Reserving the consideration of alternating-current machines for a later chapter and confining our attention for the present to the direct-current ones, we find that these latter can be further subdivided into generators, motors, dynamotors, motor generators, and boosters.

In order to present the modern direct-current types in some logical sequence, they will be taken up under these subdivisions, and any special construction, advantage, or feature will be pointed out. The machines described will be selected as illustrating the best standard practice, but it must be remembered that this treatment can not pretend to do more than introduce the reader to the subject.

GENERATORS

Capacities. Direct-current generators are manufactured in standard sizes, varying in capacity from a fraction of a kilowatt to 2700 kilowatts. Sizes above 200 kw. are nearly always direct-driven, although belted generators are found up to 500 kw. Below 200 kw. either method is employed, depending upon the special conditions of the case.

Speeds. The speeds at which direct-current generators are driven, vary from 70 or 75 revolutions per minute in the largest sizes of engine type, up to 2000 in the belted and 3000 in the steam-turbine driven types. In general, the greater the kilowatt capacity the lower the speed, although even this general rule does not hold in comparing different lines. Belted generators are very seldom run below 300 to 400 revolutions per minute, while 325 is a rather high speed for a direct-driven generator, if we except the generating sets and marine sets run by vertical engines. These latter, in the very smallest sizes, reach a speed of 850 revolutions per minute.

Number of Poles. Modern direct-current generators are always multipolar as soon as their capacity is above 5 or even 3 kilowatts, the number of poles increasing with the size of the generator. Belted generators usually have 4, 6, or 8, and occasionally 10 poles. Direct-driven generators, since the speed is fixed by the prime mover, are designed to meet these conditions. Such machines seldom have less than 6 poles and in the larger sizes and slower speeds very often have from 24 to 36.

CLASSIFICATION

Direct-current generators, according to their electrical design, may be divided into two main classes: those giving constant current, feeding series circuits; and those maintaining constant voltage, feeding parallel circuits.

Constant Current Types. Constant current generators are still found supplying current to series street arc-lighting circuits, and in the types used at present deliver approximately from 5 to 10 amperes, their voltage being variable, depending upon that needed by the circuit they are supplying. The manufacture of these series generators has been practically discontinued. New installations for street lighting employ alternating current or else direct current obtained by means of mercury arc rectifiers. The series machines still in use are of the Brush, Thomson-Houston, and Wood types. Of these, the Brush and the Thomson-Houston have open coil armatures, and the Wood has a closed coil armature. As these machines are rapidly becoming obsolete, no further description will be given of them.

Constant Voltage Types. Constant voltage generators may be further subdivided into those giving 5 to 12 volts, those giving about 125 volts, those giving about 250 volts, those giving 500 to 750 volts, and those giving 1200 volts. Generators designed to give from 5 to 12 volts are employed for electroplating and electrochemical purposes. Where a sufficient number of vats can be connected in series, however, regular standard 125-volt machines are preferably employed. Generators giving 125 volts are used for indoor arc and incandescent lighting, for the running of smaller motors, and in connection with the three-wire system. Machines giving 250 volts are used principally for power purposes, and, with auxiliary devices or parts, can supply a three-wire system. Machines giving 500 to 750 volts are employed as generators for power and railway circuits, while 1200-volt generators are the most recent development in higher voltage direct-current railway installations. All railway generators, although denoted as constant voltage, are generally designed so that, by means of their compound winding, they over-compound, usually 5 or 10 per cent; that is, at full load their voltage is 5 or 10 per cent higher than at no load.

The uses to which direct-current generators are put, we find,

Fig. 17. Allis-Chalmers Type I Engine Generator
Allis-Chalmers Manufacturing Company

therefore, to be in connection with series arc lighting, constant potential lighting, power, railway, and electrochemical purposes.

DESCRIPTION OF TYPES

Allis-Chalmers Manufacturing Company. *Engine Types.* This company has designed a line of engine type generators for general lighting and power service that are built for standard full load pressures of 120, 240, and 525 volts. They are compound-wound, giving variation in voltage from no load to full load; the no-load voltages being 115, 230, and 500, respectively. By proper increase in speed, these machines also operate satisfactorily at 125, 240, and 550 volts, full load. They are made in a great variety of capacities and speeds, ranging from 10 to 1200 kw., and run at speeds from 90 to 470 revolutions per minute.

Fig. 17 shows a 200-kw. 240-volt 125-r. p. m. machine of this

Fig. 18. Assembled Type I Field Yoke, Poles, and Coils
Courtesy of Allis-Chalmers Manufacturing Company

type. The cast-iron field frame is of circular form; the two parts of the yoke, divided horizontally, are bolted together, as shown. In the larger sizes the upper and lower halves are bolted together by cap screws having their heads set in pockets in the lower half of the yoke. The field poles are made of steel punchings riveted together, except in some of the larger frames where forced steel is used for

GENERATOR DIRECT-CONNECTED TO BALL HORIZONTAL, SINGLE CYLINDER CORLISS ENGINE
Courtesy of Ball Engine Company

round poles. The poles are bolted to the inside surface of the yoke. This method of construction enables accurate spacing to be obtained, and, by removal of one or more poles, field coils can be changed, or minor armature repairs made, without dismantling the whole machine. The pole faces are carefully designed to give a well-graded flux distribution necessary for good commutation.

Fig. 18 is a view of assembled field yoke, poles, and coils. The field coils are ventilated by being built up in sections with air spaces between the pole and the coil and also between sections of the same coil. The coil is insulated from the pole by spacers and firmly held in place by fiber insulated pins driven into the pole sides. By the fan action of the armature air is forced through these spaces, rapidly dissipating the heat generated in the coils. This method of construction is shown by Fig. 19.

Fig. 19. Details of Type I Poles and Coils
Courtesy of Allis-Chalmers Manufacturing Company

The coils are impregnated at a high temperature with a compound rendering them moistureproof and oilproof. The armature is built up of high grade steel laminations, well japanned, containing ample radial ventilating ducts and mounted upon a spider of ample mechanical strength, so constructed as to assist in a free circulation of air about the machine. The armature coils are form-wound with ample clearance between the end connections to permit free air circulation. This feature of coil clearance is evident from Fig. 20. The armature coils are insulated with the highest grade of insulating material, then dipped and baked at a high temperature, and finally placed in a steam heated screw press and pressed into shape to fit the armature slot. This renders a coil practically unaffected by oil or moisture. The coils are held in position in the armature slots by hardwood wedges fitted into grooves at the top of the slots. No band wires are used under the poles, but a steel wire band at each end of the armature serves to keep the ends of the coils in place. This type of armature, with conductors placed in deep slots, and not wound on the armature surface, is known as an *ironclad armature*, and is the standard form adopted by all manufacturers today.

This line of machines conforms to the standard practice of cross-connecting all multiple-wound armatures at points of equal potential to compensate for any inequality of the magnetic circuit. (See "General Electric Company", Fig. 48, showing equalizer rings.) The commutator is constructed of hard drawn copper bars separated

Fig. 20. Armature and Commutator of Allis-Chalmers Generator

by high grade mica of such hardness as to wear uniformly with the bars. The brush yoke, as seen from Figs. 17 and 18, is supported on the frame by three small rollers and is designed to secure a rigid structure and to leave the brushes and commutator easily accessible. The position of the brushes is adjusted by turning the handwheel.

Standard Three-Wire Type. The Allis-Chalmers Company builds a line of standard three-wire generators for output from 35 kw. to 250 kw., using their engine type frames for this purpose. These machines are wound for 240 volts between line wires and 120 volts between each line wire and the neutral. (See "Crocker-Wheeler

Company" and "Westinghouse Electric and Manufacturing Company".)

Belted Types. The belted generators put upon the market by the Allis-Chalmers Company are intended to fulfill every requirement except those for the smallest capacities. The materials, methods, and processes employed are the same as for the engine type. The large sizes have a third or outboard bearing, this being accomplished like the construction shown in Fig. 45. For the

Fig. 21. C & C Type G Direct-Connected Generator
Courtesy of C & C Electric & Manufacturing Company

smaller belted sizes this company uses its type K machines. These latter have the same frames as the company's motors of this line, becoming compound generators by merely changing the shunt field winding and adding a series winding. These frames will be described in detail later, therefore, under "Motors".

C & C Electric & Manufacturing Company. *Direct-Connected.* The direct-connected type G machines of this company furnish a line of generators overcompounded 5 per cent and are wound for voltages of 125, 250, or 500. They are built in sizes from 25 to 500 kw. in both two- and three-wire types. A 250-volt 300-kw. machine is shown in Fig. 21. The magnet frame is circular and, together with the poles, is of soft cast steel of high permeability. It is divided horizontally. The poles are round, to give minimum field copper,

Fig. 22. 10-Pole Field Ring Showing Shunt and Series Coils
Courtesy of C & C Electric & Manufacturing Company

and are bolted to the yoke, thus being readily removed. They are specially slotted so as to divide them into sections, thereby having an air space through their center and parallel to the shaft. This increases the flux in the pole tips and materially decreases the armature reaction. The field coils are wound upon non-combustible insulated bobbins, the series and shunt coils being wound separately and carefully insulated. The series coils consist of a continuous laminated copper conductor joined only between the field spools by heavy screws. These details are illustrated in Fig. 22.

The armature spider is of substantial construction, its hub forming a long sleeve key seated on the shaft. The laminations of selected sheet steel, doubly annealed and thoroughly insulated from

Fig. 23. C & C Type G Armature

each other by varnish, are securely keyed to the spider. The periphery of the core is slotted for receiving the armature coils, which are form-wound copper bars and interchangeable. They are continuous from end to end, having no soldered joints, except where connected to the commutator risers, and they are retained in place by wedges, or band wires, or both. In multiple-wound armatures, equalizing rings or cross connectors are employed. The commutator spider is mounted on an extension of the armature spider to which it is keyed, thus allowing the shaft to be removed without disturbing the commutator. The commutator bars are of the best hard drawn copper of highest conductivity, insulated from each other and from the shell with the best grade mica. A self-contained armature and commutator of this line of machines is shown in Fig. 23. A polished series shunt of non-flexible

Fig. 24. C & C Adjustable Series Shunt

Fig. 25. C & C Type G Three-Wire Generator
Courtesy of C & C Electric & Manufacturing Company

Fig. 26. C & C Type M P L Belted Generator
Courtesy C & C Electric & Manufacturing Company

grids with adjusting screws is employed for adjusting the compounding to any desired degree. Such a shunt is illustrated in Fig. 24.

Three-Wire Types. In C & C three-wire generators, well-ventilated auxiliary or balancing coils are attached to the armature spider on the side away from the commutator of a standard two-wire engine type generator. The ends of these coils are tapped into the armature winding, and the neutral point is taken from a single collector ring mounted on the commutator ring from which it is insulated. A ring at the end of the brush holder studs supports two additional studs for the brush making contact with the collector ring. Such a machine is shown in Fig. 25.

Fig. 27. C & C Type SL Belted Generator

Belted Types. A line of belted machines designated MPL is illustrated in Fig. 26. They embody the same features as the engine type and the same methods of manufacture are employed in their construction. In the larger sizes they are three-bearing.

Another line of belt-driven machines is illustrated in Fig. 27 These machines, designated type SL, are all of the 4-pole type. The magnet frame or yoke, the four poles, and the supporting feet are cast of the best soft steel in one piece. The protecting rings, which protect the field coils and support the brush rigging, are cast integrally with the bearing brackets. This piece is of cast iron and is attached to the magnet frame by dowel pins and cap screws. The field coils are carefully wound on easily removable sheet iron bobbins. All C & C machines employ special brush holders of the reaction type, carrying carbon brushes inclined at a slight angle from the radial line, and also use bearings of the self-oiling self-aligning type in which brass rings in the oil wells carry oil to the shaft. The generating sets put upon the market by this company are generators selected from the preceding lines, directly coupled to vertical high speed engines of various other manufacturers, an example of which is shown in Fig. 28.

Crocker-Wheeler Company. *Engine Types.* The line of engine type generators built by this company is in sizes from 200 to 1500 kw. rated at 125 or 250 volts, running at various speeds from 200 to 80 r. p. m., and having from 10 to 16 poles. For direct connection they also build a second line of 6- and 8-pole machines in sizes from 25 to 250 kw., whose speeds range from 325 to 150. One of the larger sizes is shown in Fig. 29. In these generators the magnet frame is of cast iron proportioned to insure ample stiffness in the

Fig. 28. C & C Type S L Generator Direct-Connected to High Speed Engine

smaller line, while in the larger it is stiffened by internally projecting flanges. The frame is split horizontally, the two halves being accurately aligned by dowel pins and bolted together. The lower half of the frame is provided with feet bolted down to the supporting base and supplied with leveling screws for accurately adjusting the position of the magnet frame. The poles are of steel securely bolted to the frame or, in the smaller line, cast welded into the frame. This is accomplished by constructing the poles first and placing them in position in the mould in which the magnet frame is to be cast. In casting the magnet frame, the molten metal flows around the ends of the poles so that, on cooling, one solid mass results. If the cast welded joint is perfectly made, it results in a better magnetic circuit than is secured by the method of bolting in the poles. Each pole is fitted with a cast-iron removable shoe that produces the desired flux distribution and also holds the field coils in position.

Fig. 29. Crocker-Wheeler 1200-K. W. Engine Type Generator
Courtesy of Crocker-Wheeler Company

For some of the ratings of these generators the use of commutating poles has been found desirable. These are of solid steel, rectangular in cross section and carrying lugs on the ends to support their windings. The purpose of the commutating poles is to compensate armature reaction at all loads and secure excellent commutation. (See "Electro-Dynamic Company", page 68.) The air gap or clearance between the armature and the pole face is relatively large. This is a good feature, as it tends to reduce the bad effects occasioned when the armature is slightly out of center. The field coils on each pole are divided into sections separated from each other and from the pole core by spacers so as to provide increased ventila-

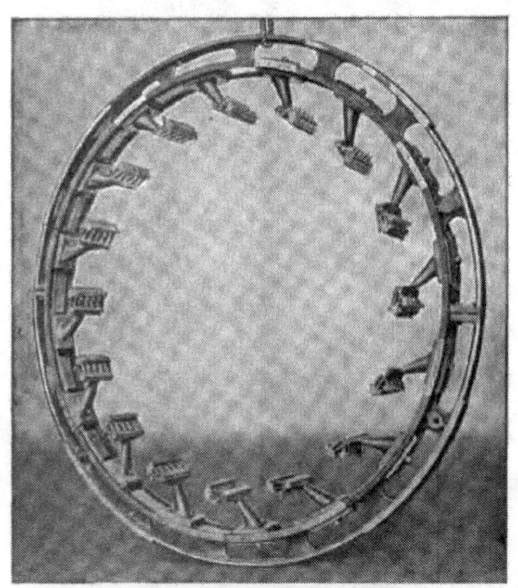

Fig. 30. Brush Rigging for 600-K. W. Generator
Courtesy of Crocker-Wheeler Company

tion. The windings of the series coils and also the commutating pole windings are of strip copper wound on edge. The connections of these windings between the various coils are made by interleaving the multiple strips, thus securing the best mechanical and electrical contact. The armature spider, consisting of a hub with projecting arms, supports the toothed laminations of the armature core. The armature conductors consist of flat or round copper wire without joints of any kind. They are thoroughly insulated and are retained in the slots by wedges or band wires. The commutator spider is mounted on an extension of the armature spider, thus being independent of the shaft.

The cast-iron rocker ring carrying the brush rigging is rigidly supported from the magnet frame and has means for shifting simultaneously the position of all the brushes. All the positive sets of brushes are connected to a copper brush ring mounted on one side

Fig. 31. Crocker-Wheeler Parallel Movement Brush Holder

of the rocker ring, and all the negative brushes to another ring similarly mounted on the other side of the rocker ring. A positive low resistance connection is provided between each brush and its brush holder bracket. The brushes are held by parallel movement holders that are individually adjustable and will not allow the position of the brush to change as it wears away. These features are shown in Figs. 30 and 31. The Crocker-Wheeler Company also manufactures a line of 550-volt engine type railway generators embodying the preceding features, in sizes from 150 to 1500 kw., with 8 to 16 poles, and running at 275 to 80 r. p. m.

Belted Types. Crocker-Wheeler belted generators are built in two separate lines, their form H and their form I machines. Form H generators, furnishing the sizes above 45 kw., employ the same

Fig. 32. Crocker-Wheeler Form I Generator

Fig. 33. Crocker-Wheeler Generator Direct-Connected to American Blower Company Engine
Courtesy of Crocker-Wheeler Company

frames as the commutating pole motors of this name, and the features of this frame will be given under "Motors". Their form I machines, shown in Fig. 32, give them the smaller sizes from 3 to 45 kw., ranging in speed from 1500 to 760 r. p. m. These machines are multipolar and have a round cast-iron frame of which the feet form a part. The bearings are supported by four-arm castings attached to the yoke. The poles are of steel cast welded into the frame. The armature coils are of heavily insulated wire and form-wound.

Fig. 34. Crocker-Wheeler Three-Wire Generator

After being taped and dipped in an insulating varnish they are baked.

Miscellaneous Types. These machines, together with Crocker-Wheeler form L motor frames for the smaller sizes, when direct-connected to vertical high-speed engines of other manufacturers, give rise to generating sets from $1\frac{3}{4}$ to 25 kw., running from 850 to 325 r. p. m. These are illustrated in Fig. 33. The three-wire generators put upon the market by this company are in most respects identical with their standard two-wire machines. One of them is shown in Fig. 34.

The series field windings are divided into two equal sections. The windings on the positive poles are connected to one side of the armature circuit, and the windings on the negative poles to the other side. The armature slots are cut deeper than ordinarily, to provide room for auxiliary windings in the bottom of the slots. The type of auxiliary winding shown diagrammatically in Fig. 35 is known as the polyphase winding and consists of several windings, usually more than three, similarly connected to the armature with the neutral

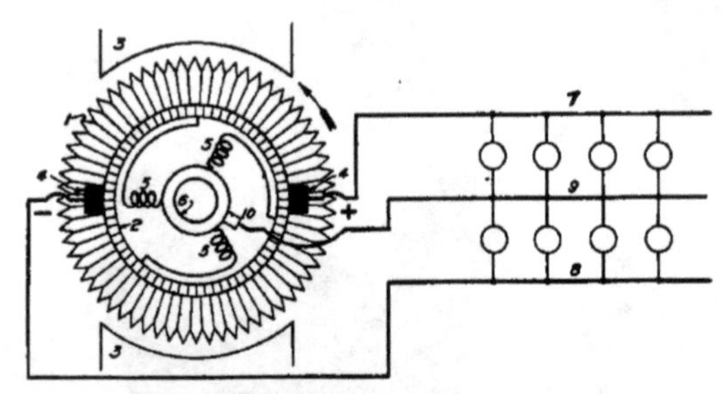

Fig. 35. Diagram of Auxiliary Windings of Crocker-Wheeler Armature. 1, Armature Winding; 2, Commutator; 3, Poles; 4, Brushes; 5, Auxiliary Winding; 6, Slip Ring; 7, Positive Wire; 8, Negative Wire; 9, Neutral Wire

point connected to the slip or collector ring. The polyphase winding is so distributed over the face of the armature that the average field in which it moves is uniform at all times. Any tendency to flicker is thereby entirely overcome. The collector ring is mounted at the outer end of the commutator and supported from the commutator spider.

Fairbanks, Morse & Company. *Engine Types.* The engine type generators manufactured by this company are illustrated in Fig. 36, showing a 200-kw. 250-volt machine. They are made in sizes from 50 to 300 kw., with speeds from 275 to 100 r. p. m. The frame or magnet yoke is made of special dynamo iron of high permeability. The pole cores are of extra quality re-hammered wrought iron, circular in cross section and bolted to the frame. The pole shoes are of laminated sheet steel, so shaped as to give the correct

Fig. 36. Fairbanks-Morse Engine Type Generator

Fig. 37. Fairbanks-Morse Belted Generator

flux distribution and avoid eddy currents. The field coils are wound on metallic spools, the series and shunt coils being in separate com-

Fig. 38. Large Type of Fairbanks-Morse Belted Generator

partments. The armatures are ironclad, bar-wound with one piece coil, and well-ventilated.

Belted Types. This company also manufactures several lines of

Fig. 39. Fairbanks-Morse Field Coils and Pole Piece

moderate and slow speed belted generators. Their moderate speed machines range from 2 to 125 kw., running 1850 to 675 r. p. m.

2500-K.W. 3-PHASE 2300-VOLT WATER-WHEEL-DRIVEN ALTERNATORS AND EXCITERS, CONNECTICUT RIVER POWER COMPANY, VERNON, VERMONT

Courtesy of General Electric Company

These same machines, when run at lower revolutions per minute and suitably changed as to windings, become their slow speed generators rated from 1½ to 100 kw., running at 1800 to 550 r. p. m. The smaller sizes are illustrated in Fig. 37, while the larger take the form shown in Fig. 38. In these lines the frame is a semi-steel casting. The poles are built up of steel laminations pressed solidly together between steel side pieces, or in the smaller sizes cast solid with the frame. Pole horns support the field coils and give the proper flux distribution in the air gap. The shunt field coils are wound upon cast-iron forms and are thoroughly insulated. The series coils are made of strip copper, insulated with mica, and protected with a cord winding on the outside. This construction is shown in Fig. 39. In the smaller machines the armature coils are form-wound.

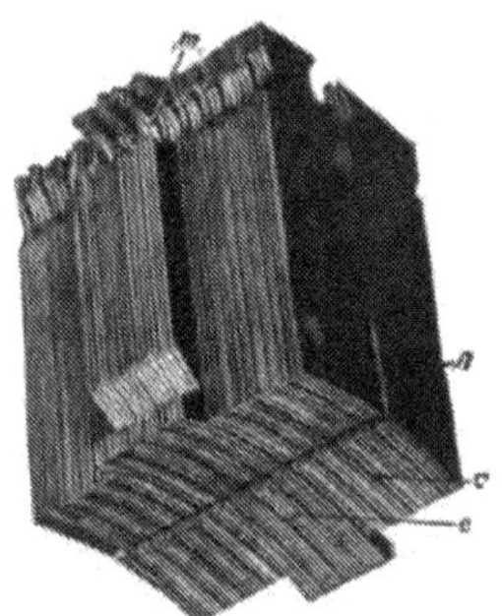

Fig. 40. Fort Wayne Laminated Pole Piece

Fort Wayne Electric Works. *Engine Types.* Although this has become a subsidized company of the General Electric Company, their apparatus retains its distinctive features. The frames of the engine type generators are made in two styles, those divided horizontally and those divided vertically. The former are made in sizes from 15 kw. to 1000 kw., and the latter from 50 kw. to 1000 kw., inclusive. The vertically divided frames, by means of proper screw mechanism, can have the two halves moved apart horizontally in a direction at right angles to the shaft. The pole pieces are made up of sheet iron punchings and sheets of insulating material, as shown at *e*

Fig. 41. Fort Wayne Ventilated Field Coils

Fig. 42. Horizontally Split Frame with Armature Removed
Courtesy of Fort Wayne Electric Works

Fig. 43. Fort Wayne Armature Partly Wound

and *c*, Fig. 40. These are riveted together, slotted to about one-half their length, as shown at *n*, and provided with grooves *m* for holding them securely in the frame. After being cast welded into the frame, they are machined and, in the larger sizes, fitted with pole shoes. The field coils are wound on insulated metal forms, as shown in Fig. 41, providing good ventilation. Fig. 42 shows a horizontally split frame with armature removed and clearly illustrates

Fig. 44. One-Piece Frame of Type L F Generator
Courtesy of Fort Wayne Electric Works

these details, including the method of shifting the brush yoke. The armature is of the usual ironclad, one-piece coil, well-ventilated type. A partly wound armature is shown in Fig. 43.

Belt-Driven Types. Fort Wayne belt-driven types are manufactured in sizes from $\frac{3}{4}$ to 400 kw. They embody the same charac-

Fig. 45. Fort Wayne 3-Bearing Belted Generator

Fig. 46. General Electric Railway Generator

teristics as the direct-connected type, but the base, field frame, pulley, and pedestal, and part of the commutator end pedestal, are all cast in one piece, with the laminated pole pieces cast welded into the yoke. Fig. 44 clearly shows this method of construction and Fig. 45 shows one of the larger size three-bearing belted generators.

General Electric Company. *Railway Types.* Fig. 46 shows a direct-connected railway generator built by this company. These machines range in size from 100 to 2700 kw., having from 6 to 26 poles and running at 275 to 75 r. p. m. They are compound-wound, generally furnishing 525 volts at no load and 575 at full load. This company's line of direct-driven lighting and power generators is of similar design, from 25 to 200 kw., 6- and 8-pole, running at 310 to 110 r. p. m. and giving 125 or 250 volts. Sizes giving 275 volts are from 300 to 2000 kw., with 10 to 24 poles, running at 150 to 100 r. p. m.

Fig. 47. General Electric Ventilated Field Coil

In all these machines the external circular yoke of the field is made of cast iron of oval cross section, except in the very large sizes where it is cast with a box-like construction to give greater stiffness. The two halves are fastened together by bolts entirely hidden. The poles are solid steel castings, accurately fitted and bolted to the yoke. They are also keyed in such a way that they may be slipped out laterally. Pole shoes, laminated in machines above 200 kw., are designed to give a graduated flux density in the air gap and also serve to keep the field coils in place. The field coils are rectangular in shape, as shown in Fig. 47. The end flanges of the spools are ventilated and ducts are formed in the coils.

In the larger machines, the series and the shunt coils are wound on separate parts of the bobbin, while in the smaller sizes the shunt winding is over the series. The toothed armature core laminations are built up upon the rim of a strong central spider shown in Fig. 48. At suitable intervals between laminations, space blocks, as illustrated in Fig. 49, are inserted. These are made of steel strips set on edge and interlocked with the laminations, thus forming ventilating

Fig. 48. Armature Showing Equalizer Ring
Courtesy of General Electric Company

ducts or air passages. The arms of the spider have cast upon them wings or fan blades. These serve as a powerful centrifugal fan to

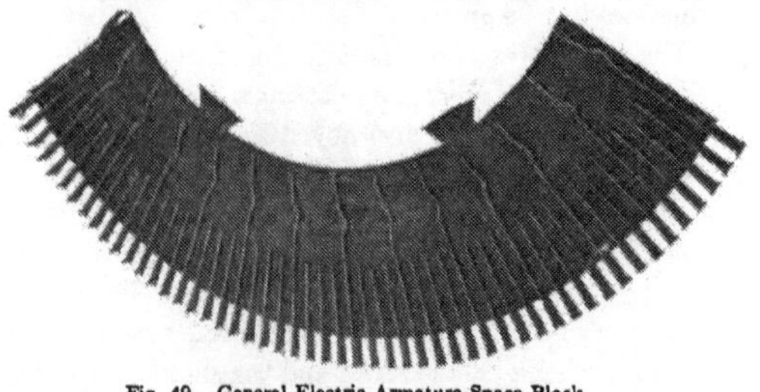

Fig. 49. General Electric Armature Space Block

keep a constant blást of air passing between the laminations and windings and around the poles. An armature embodying these features is called a *ventilated armature.*

The armature coils are of the form-wound copper strip type, held in the slots by wooden wedges. Equalizing rings are provided on generators having multiple-wound armatures. These rings, mounted on the end flange on the back of the armature, as shown in Fig. 48, are used to connect the armature windings between points of equal potential, so that any unbalancing that may occur will be equalized by the alternating currents that will flow through these rings between the sections. These currents, due to the armature reaction, equalize the pole strengths, thus causing an equal division of the direct current in the several paths, and they thereby improve

Fig. 50. Field Frame Showing Compensated Windings
Courtesy of General Electric Company

the commutation very much. Commutators and brush holders are of generous design, allowing large radiating and contact surfaces. In the General Electric railway generators all sizes above 1000 and below 400 kw. are equipped with commutating poles. These compensate armature reaction at all loads and secure excellent commutation. (See "Electro-Dynamic Company", page 68.)

During the past few years railway generators wound for potentials of 1200 to 1500 volts and capable of carrying three times normal load, have been developed. Owing to the necessity of carrying such heavy overloads, it is usual to provide, in addition to the commutating field, a compensating winding so proportioned as to practically nullify the effect of armature reaction. This winding consisting of heavy copper bars, is placed in slots in the pole surface and connected in series with the armature, commutating field, and series field. It will be readily seen that the current flowing in these windings will rise and fall at the same rate as the current in the armature winding, thereby preventing flux distortion and resultant sparking. In Fig. 50 can be seen the compensated windings for a 500-kw. field frame.

Fig. 51. General Electric Belted Generator on Cast-Iron Sub-Base

Belted Types. The General Electric Company builds several lines of belted machines. Fig. 51 illustrates their medium size belted machine. These are 6-pole, 16- to 150-kw. capacity, and running at speeds from 1100 to 500 r. p. m. They are wound for 125, 250, or 500 volts. Another line, called type D L C, employs commutating poles. These range in capacity from 20 to 150 kw., run at speeds from 1425 to 650 r. p. m., are 4- or 6-pole, and are wound for standard voltages of 125, 250, and 575. The main poles are made of laminated iron, while the commutating poles are of smaller section and are constructed of machine steel. The special features are clearly shown in Fig. 52.

Fig. 52. Armature Magnet Frame and End Shield of D L C Generator
Courtesy of General Electric Company

Three-Wire Types. General Electric 250-volt generators become three-wire generators by adding two collector rings, mounted at

Fig. 53. General Electric Three-Wire Generator

the commutator end and connected to the proper points in the armature winding. Machines of this type up to 200 kw. are illustrated in Fig. 53. The brush rigging for the collector rings is supported from the pillow block and makes connection by copper brushes to a compensator. These compensators are similar in construction to transformers and consist of two or more insulated coils wound on a laminated iron core. They carry an alternating current and serve to maintain a neutral potential at their middle point, to which, there-

Fig. 54. Generating Set with Forced Lubrication Engine
Courtesy of General Electric Company

fore, can be connected the middle or neutral wire of a three-wire system.

Generating Sets. This company also manufactures a line of small direct-current generating sets, both generator and engine complete, in sizes from 2½ to 75 kw., running at speeds from 700 to 280 r. p. m. In the smaller sizes the generators are 4-pole, 110-volt, while in the larger they are 6-pole, 125-volt. Designed primarily to meet the severe conditions of marine work, which demands light, very compact, and extremely durable sets of close regulation, these generators are in addition employed for power and lighting in isolated plants and as exciters in central stations. Fig. 54 illustrates a 50-kw. set of this type.

Turbine-Driven Types. The direct-current Curtis steam tur-

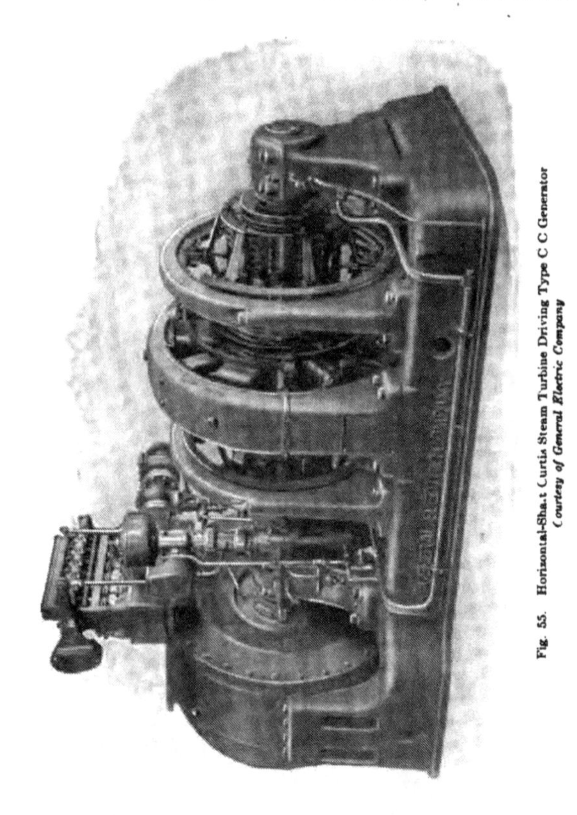

Fig. 55. Horizontal-Shaft Curtis Steam Turbine Driving Type C C Generator
Courtesy of General Electric Company

bine sets brought out by the General Electric Company are arranged with horizontal shaft as illustrated in Fig. 55. They are built in sizes from 5 to 300 kw., at 120 and 250 volts. The generators are of the most recent and improved types. By the use of commutating poles,

Fig. 56. General Electric 1000-Ampere Electrolytic Generator

Fig. 57. Holtzer-Cabot Belted Generator

sparking under all conditions of load is eliminated. Special commutator constructions on account of the high speeds and massive brush rigging are also employed. The generators are driven by overhung turbines consisting of one or two wheels, thus keeping the bearings down to two in number.

Low Voltage Types. This company also builds low voltage generators to be used for electrolytic work. These range from 1½ to 15 kw., running from 1800 to 1200 r. p. m., and giving from 2 to 10 volts. They are double commutator machines, as shown in Fig. 56, the smaller ones being self-exciting, while the larger employ small high-speed bipolar series exciters.

Fig. 58. Holtzer-Cabot 1000-Ampere Plating Generator

Holtzer-Cabot Electric Company. Fig. 57 shows the belted generators built by this company. The line of these machines includes sizes from ¾ to 120 kw. at speeds from 1800 to 500 r. p. m. The field ring in the smaller sizes is a one-piece casting, but it is split in the larger sizes. The poles are of wrought iron, cylindrical, and cast welded into the yoke. The field coils are form-wound with mica reinforced insulation. The armature is ironclad with windings of round, ribbon, or bar copper depending upon the size of the machine. The armature coils are one piece and held in the slots by maple strips and non-magnetic binding wires. The commutator segments are drop-forged copper in the smaller sizes and hard drawn copper in the larger sizes. The shaft is of crucible steel with the bearings of hard phosphor bronze, self-aligning and self-oiling, while the pedestals are bolted to the base. By changing the windings and

Fig. 59. 200-Kilowatt Engine Type Generator
Courtesy of Ridgway Dynamo and Engine Company

Fig. 60. Ridgway 150-Kilowatt Belted Generator

modifying the commutator and brush constructions, these frames are adapted to plating generators, as shown in Fig. 58.

Ridgway Dynamo and Engine Company. This company manufactures complete engine-driven units and also belted generators from 10 to 750 kw., in the three standard voltages of 125, 250, and 550, illustrated in Figs. 59 and 60. These machines in addition to employing the usual methods and materials of high grade generator construction, have several distinguishing features. The field ring or yoke is constructed of steel laminations, the punchings being securely

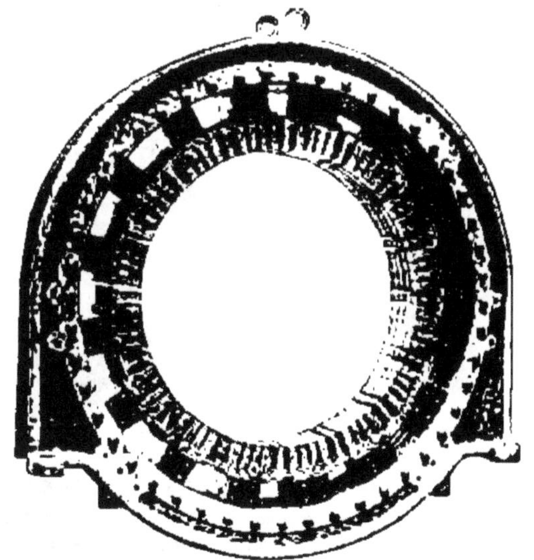

Fig. 61. Ridgway 400-Kilowatt Field Ready to Wind

held between heavy cast-iron clamping rings having a modified I-beam section. The pole pieces are also laminated and are built up in two separate parts. One part forms the core for the shunt field coil, while the other is the pole shoe. The two parts are firmly bolted to the field ring by heavy cap bolts that pass through the pole core and screw into the pole shoe.

Between the pole shoes are placed commutating poles, or interpoles. These poles also are built of laminated steel, and are supported from the pole shoes by brass keys driven into slots in the sides of the poles themselves and the adjacent pole cores. The pole shoes are provided with three slots each, through which are wound the "balancing coils". These coils take the place of the series field coils of the ordinary compound generator. Balancing coils lie parallel to the armature conductors and are so wound that the current in them flows in the opposite direction to that in the armature conductors. The coils are connected in series with the armature and thus are enabled to set up a local magnetic field, opposite in direction and just

Fig. 62. Sturtevant 8-Pole Direct-Driven Generator

balancing that of the armature winding. This holds the field, due to the shunt coils, in the same place for all loads, and gives a fixed plane of commutation. The central portion of each balancing coil is wound around the commutating pole and sets up a commutating field, giving sparkless commutation with fixed brush position at any and all loads. Winding the balancing coils eccentrically and also adding a few extra turns produce a compounding effect. Fig. 61 illustrates the laminated steel field construction here described.

These machines become three-wire generators by the addition of two collector rings furnishing alternating voltage to a choke coil whose middle point is connected to the neutral wire of the system. The balancing coils are now divided and half are connected on one side and the remainder on the other side of the armature.

B. F. Sturtevant Company. This company manufactures both the steam engines and the generators of their direct-driven units. These lines of units comprise an engine type generator in capacities from 20 kw. at 375 r. p. m. up to 150 kw. at 250 r. p. m., all 8-pole, driven by horizontal steam engines. Sturtevant 10-pole generators range from 150 kw. at 200 r. p. m. to 500 kw. at 160 r. p. m., these latter being driven by vertical compound engines.

Fig. 63. Sturtevant 6-Pole 5-Kilowatt Generating Unit

This company's generating sets of 6- and 8-pole generators combined with various vertical engines give capacities from 3 kw. at 600 r. p. m. up to 100 kw. at 350 r. p. m. Figs. 62 and 63 illustrate these types. The magnet frame is of cast iron split horizontally. The pole pieces are through bolted to the yoke and carry cast-iron shoes. The field coils are machine-wound of open construction to secure maximum ventilation, the series and shunt coils being wound separately. The armature is of the ironclad type with the steel laminations mounted on a cast-iron spider. The armature coils are made waterproof and oilproof by dipping in armalac and baking for 24 hours in a temperature of 100°C. Within the armature core are space blocks having radial arms that act like the blades or the vanes of a centrifugal blower, thus increasing the ventilation.

This company also puts on the market a line of steam turbine generator sets ranging from 3 kw. at 3000 r. p. m. to 75 kw. at 2000 r. p. m., the generators being wound for voltages from 100 to 250. One of these is illustrated in Fig. 64.

Fig. 64. Sturtevant Steam Turbine Direct-Connected to Generator

Triumph Electric Company. *Engine Types.* Fig. 65 shows one of a line of Triumph engine type generators, ranging in size from 30 to 1000 kw. The magnet frame is made of close grained cast steel with pole pieces of the laminated type bolted to the frame. Sizes smaller than 250 kw. have the frame split vertically, while the larger sizes have the frame split horizontally. The shunt and series coils are wound separately, the series coil being formed of solid bar copper, with spaces left between the coils. The armature is of the self-contained ironclad type. With the exception of the higher voltage and the smaller machines the armatures are bar-

Fig. 65. Direct-Connected Triumph Generator
Courtesy of Triumph Electric Company

Fig. 63. Triumph Generator Direct-Connected to Troy Engine

Fig. 67. Triumph Belted Generator

wound, each coil consisting of one continuous length of solid copper, and held in position by hardwood wedges.

Generating Sets. This company also builds a line of generating sets of the marine type as shown in Fig. 66, in sizes from 4 to 25 kw., all 4-pole machines and running at 500 to 350 r. p. m.

Belted Types. Triumph belted generators are built in sizes ranging from ½ to 100 kw. The frame is made of close grained

Fig. 68. Three-Wire Belted Generator
Courtesy of Triumph Electric Company

steel. The larger sizes are built with laminated poles and the smaller sizes employ solid steel poles with laminated tip or shoe. The larger frames employ a three-arm bracket, as shown in Fig. 67, to support the bearings, while in the smaller frames only a two-arm bracket is employed.

Three-Wire Types. This company's line of three-wire generators is illustrated in Fig. 68. These machines are built in all standard sizes from 25 kw. up and are wound for 250 volts. They differ from the single voltage machines only by the addition of three collector rings tapped into the direct-current armature winding at the proper points. These collector rings are attached to three reactance coils, connected in star, their neutral point being joined to the neutral wire of the three-wire system.

Westinghouse Electric and Manufacturing Company. *Interpole Type.* This company builds a line of direct-connected generators. designated as type Q, in sizes from 25 to 1000 kw. Up to 100 kw. they are wound for 125 or 250 volts at full load; from 100 up to 300 kw. they are wound for either 125, 250, or 600 volts at full

Fig. 69. Westinghouse Type Q Main Coil Windings

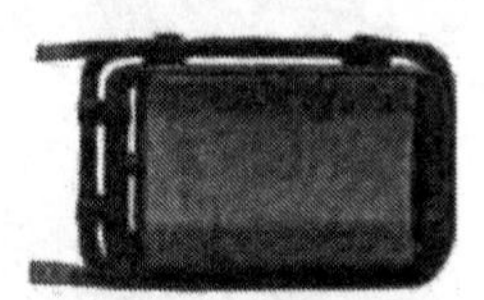

Fig. 70. Westinghouse Type Q Field Coils Showing Air Space

load; and above 300 kw. they are wound for 250 or 600 volts at full load. They are all compound-wound and employ commutating poles or interpoles. The no-load voltages are 118, 230, and 550. The field frames are of cast steel of approximately elliptical cross section. Machines having a capacity of 50 kw. and greater have their frames divided horizontally, while machines of small capacity have their frames cast in one piece. The pole pieces are laminated and through bolted into the frame. The main field coils are constructed so as to afford the best ventilation. An air space is provided between the inside of the shunt coil spools and the sides of the poles. The series coils are made of bare edge-wound strap windings. There is an air space between adjacent turns and between each shunt and series coil, permitting a free circulation of air about each conductor. This construction is shown in Figs. 69 and 70. The commutating poles are made of one piece of steel and are firmly bolted to the frame. The commutating pole coils are of edge-wound copper strap. The armature and commutator are constructed according to the usual Westinghouse types. Since the brushes of the interpole generators should not be shifted after being once properly located, the shifting gear is eliminated. The brush gear is supported by a cast-iron rocker ring fitted in a recess in the steel frame and rigidly held in place by cap screws and washers. Fig. 71 shows the open construction of the field coils and armature of this line of machines.

Fig. 71. Westinghouse Type Q Generator Showing Open Construction and Interpoles

Belted and Standard Types. The Westinghouse Company builds a line of self-contained generators in standard sizes from 100 to 300 kw. at 250 and 550 volts and from 100 to 200 kw. wound for 125 volts. They are adapted for belt driving or for direct connection where frames with bearings are desired. The pedestals are bolted to the bed plate and support self-oiling bearings that consist of cast-iron shells lined with babbitt metal and lubricated by means of rings that ride upon the shaft and dip into large reservoirs, in the pedestals, filled with oil. The rings, rotated by the motion of the shaft, continually bring up a supply of oil.

Another line of machines, designated type S, furnishes the smaller sizes. The methods of manufacture and materials for the various parts are the same as in the other lines, but the outward appearance is entirely different, since the bearings are supported by brackets of the skeleton type with radial arms and an outer ring bolted to the frame. (See motor of same type, Fig. 105.) The standard belted machines range from 2 to 85 kw. for 125 and 250 volts and from $3\frac{1}{2}$ to 75 kw. at 550 volts. When direct-connected they require special windings and the capacity of any frame changes, the output being roughly proportional to the speed of the armature.

Still another line, called type R, gives eight sizes from $\frac{3}{4}$ to $7\frac{1}{2}$ kw.; the smallest four sizes being bipolar, the others 4-pole. These are intended for belt driving, although the multipolar frames can be adapted for direct driving.

Three-Wire Types. Any engine type, any self-contained, and any type S generator of this company above 5 kw., provided the voltage is 250 or 550, may, by the addition of properly connected collector rings, become a three-wire generator. The Westinghouse Company prefers the two-phase arrangement. This necessitates four collector rings, as shown in Fig. 72. It further requires two compensators or autotransformers. These consist each of a single winding upon a laminated core and are each connected to two of the collector rings. The middle points of the autotransformers are connected together and to the neutral of the three-wire system.

Fig. 72. Armature of Westinghouse Three-Wire Generator

MOTORS

General Characteristics. Direct-current motors are either shunt, series, or compound. A shunt motor runs at practically constant speed for all loads, has a torque almost directly proportional to the armature current, and has a starting torque usually 50 to 100 per cent greater than full load running torque. A series motor runs at greatly decreasing speed for increasing loads, has a torque that increases almost as the square of the armature current or at least much faster than proportional to it, and has a powerful starting torque. A compound-wound motor approximates the shunt or the series type in its characteristics, depending upon what winding preponderates.

Speed Classification. Direct-current motors, according to their speed performance, fall into four different classes:

(1) *Constant speed motors,* in which the speed is either constant or does not materially vary, as shunt motors and compound motors in which the shunt field preponderates.

(2) *Multispeed motors* (two-speed, three-speed, etc.), which can be operated at any one of several distinct speeds, these speeds being practically independent of the load, such as motors with two armature windings.

(3) *Adjustable speed motors,* in which the speed can be varied gradually over considerable range, but when once adjusted remains practically unaffected by the load, such as shunt or slightly compounded motors designed for a considerable range of shunt field variation.

(4) *Varying speed motors,* or motors in which the speed varies with the load, decreasing when the load increases, such as series motors and compound motors in which the series winding preponderates.

DESCRIPTION OF TYPES

Allis-Chalmers Manufacturing Company. The larger constant speed motors of this company employ the same frames as their belt generators. These motors include many sizes at 110 and 220 volts from 50 to 400 horsepower, ranging in speed from 800 to 180 r. p. m. Allis-Chalmers type K machine, shown in Fig. 73, is built in different frame sizes with a number of ratings for each frame, depending upon the winding and the speed. The smallest shunt-wound constant speed size in the 110-volt line is ½ h. p. at 500 r. p. m., and the largest is 80 h. p. at 500 r. p. m. They are also wound for 220 and 500 volts. They can be used as adjustable speed motors with a range in speed of 1 to 3. They can be wound as any type compound or series motors, thus becoming varying speed motors.

The cylindrical field magnet yoke is of open hearth steel and is machined on each end to receive the housings that carry the bearings. The housings are held in place by through bolts and on 4-pole machines can be rotated 90 degrees or 180 degrees to allow side wall or ceiling mounting; bipolar machines can be arranged for floor or ceiling mounting. The pole cores fastened to the yoke by cap screws are circular in cross section and made of open hearth steel. The pole shoes are of steel punchings riveted together and fastened to the cores by screws. The field coils effectively impregnated so as to be

Fig. 73. Allis-Chalmers Type K Motor

moistureproof are wire machine-wound. The armature is ironclad, of the ventilated type with interchangeable form-wound coils.

Fig. 74 shows an Allis-Chalmers type K motor direct-connected to the work. Standard motors of this type are made open at the ends; they can, however, also be made semi-enclosed and totally enclosed by the addition of suitable metal enclosing covers fitted to the end housings. The semi-enclosed type uses perforated covers forming a screen that, while allowing a circulation of air, protects the motor from flying particles. Totally enclosed motors are dust-proof and moistureproof. This shuts off the ventilation, and the output of a given frame is less for the totally enclosed type than for

Fig. 74. Type K Motor Driving a Band Saw
Courtesy of Allis-Chalmers Manufacturing Company

the others. The enclosed motor does well, however, for intermittent work where it has ample time to cool in the intervals of work.

The Allis-Chalmers Company also manufactures railway motors, as illustrated by Fig. 75, of the 4-pole series type, fitted with interpoles. The cast-steel field frame is split horizontally through armature and axle bearings, the lower half arranged to open downwards. The main pole pieces are of soft steel punchings, securely clamped between and riveted to malleable iron end plates. The commutating poles are of solid steel. The armature is ironclad with ventilating ducts. The coils are wire-wound, a sufficient number being bound

Fig. 75. Allis-Chalmers Railway Motor with Frame Opened

and insulated together to form a coil group. The two brush holders are rigidly mounted in the top half field frame.

C & C Electric & Manufacturing Company. This company employs the frames of their line of standard belted generators, shown

Fig. 76. C & C Open Wall S L Motor

in Fig. 26, for constant speed motors of the larger sizes. These machines are wound for 125, 250, or 500 volts. For the smaller sizes

Fig. 77. C & C Semi-Enclosed S L Motor

they employ the type SL frames wound shunt or compound for constant speed. The SL frames are arranged for ceiling and wall mounting as well as with vertical shaft projecting upward or down-

Fig. 78. C & C Type S L Motor Driving Reciprocating Compressor

Fig. 79. C & C Vertical Type S L Motor

ward. All of these varieties can also be semi-enclosed or fully enclosed. With proper windings they become adjustable speed motors and are built for a speed variation of 5 to 1. In the last case they are provided with commutating poles. Figs. 76, 77, 78, and 79 are illustrations of these machines. This company also manufactures multispeed motors with two armature windings and double commutators, as shown in Fig. 80. They are used for running printing presses when equipped with series parallel method of control. In the smaller sizes this type employs the SL frames.

Fig. 80. Multispeed Double Commutator Motor for Printing Press Work
Courtesy of C & C Electric & Manufacturing Company

Crocker-Wheeler Company. The larger constant speed motors of 50-horsepower capacity and larger are obtained by this company by using the same frames as for their form H belted generators. These motors all employ commutating poles. The frame or yoke is of cast iron with poles and interpoles of steel cast welded into the yoke. The main field coils are held in place by pole shoes. The bearings are supported by four arm brackets bolted to the magnet frame. These features are illustrated in Figs. 81 and 82, while Fig. 83 clearly shows the durable armature and commutator spider construction. The company's form I machines, shown in Fig. 32, give them shunt motors from 3¾ to 50 horsepower at 110, 115, 220, 230, or 500 volts, the speeds falling between 1260 and 330 r. p. m. This type of frame can be shunt-wound, series-wound, or compounded to any degree, as well as furnishing a line of field weakening, adjustable speed motors which give speed ranges of 2 to 1, 2½ to 1, or 3 to 1, as desired. It also can be arranged for floor, side wall, or ceiling mounting or with vertical shaft.

By adding proper covers to the open type it becomes enclosed or semi-enclosed, the latter either grid or gauze. Enclosed motors are particularly adapted to operate in flour mills, woodworking shops, boiler rooms, etc., where the air is continually filled with fine par-

Fig. 81. Crocker-Wheeler Form H Commutating Pole Motor

ticles. Being almost airtight, they are of a larger frame for a given output than open ventilated motors; but with proper windings and frame they may be practically of the same efficiency as the open types. They are not well adapted for continuous running, but do admirably for intermittent service. For the smaller sizes of constant speed machines, the Crocker-Wheeler Company uses a 2-pole motor called by them form L. They are made in sizes from $\frac{1}{16}$ to 7½ horsepower for operation on 115-, 230-, or 500-volt circuits and are suited for application to all sorts of light machinery. The special

FAIRBANKS-MORSE 100 KILOWATT GENERATOR INSTALLED FOR INDIANAPOLIS GAS COMPANY, INDIANAPOLIS, INDIANA

Courtesy of Fairbanks, Morse and Company

Fig. 82. Frame and Field Coils of Commutating Pole Motor
Courtesy of Crocker-Wheeler Company

Fig. 83. Crocker-Wheeler Armature Core for Form H Motor

feature in this line is that the frame is of cast steel with the poles cast integral with the rest of the frame and carrying laminated pole shoes. The sizes from $\frac{1}{3}$ to 3 horsepower furnish, when properly wound, adjustable speed motors providing speed variations of 2 to 1, $2\frac{1}{2}$ to 1, or 3 to 1.

Fig. 84 shows one of this company's many applications. To supply the demand for traveling crane motors, they have developed a line of motors shown in Fig. 85. These are made in sizes from $1\frac{1}{2}$ to 60 horsepower operating on 115, 230, or 500 volts. They are series-wound, compact, of strong and durable construction, with rectangular frame, and are protected against dirt and moisture by enclosing covers.

Fig. 84. Traveling Hoist Driven by Crocker-Wheeler Motor

Electro-Dynamic Company. A prominent type of adjustable speed motor is that of the Electro-Dynamic Company in which speed variations as great as from 6 to 1 may be obtained by field weakening. Sparking is prevented by the employment of interpoles; that is, auxiliary poles, small compared with the main poles, are located between the latter and provided with coils connected in series with the armature. The flux of these interpoles is in direct opposition to that of the armature and gives the commutation field. Since the coils of the interpoles are connected in series with the armature, the commutation field is not affected by weakening the shunt or main motor field to obtain the increased armature speeds, and is, furthermore, proportional to the load, thus producing sparkless commutation at all loads and speeds within the limits of the design of the machine. As the action of the interpoles is reversible the motors can be run equally well in either direction.

Fig. 86 is a view of one of these motors with the bearing housings and armature removed, showing the interpoles. These motors are manufactured in sizes from ½ to 75 horsepower as 4-pole machines and in sizes from 40 to 150 horsepower as 6-pole machines, operating

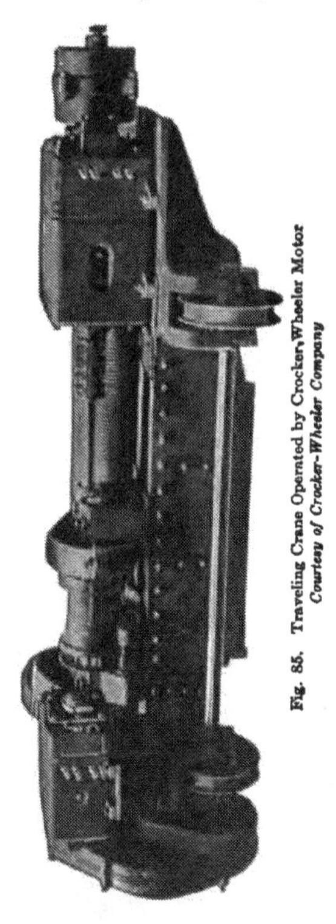

Fig. 85. Traveling Crane Operated by Crocker-Wheeler Motor
Courtesy of Crocker-Wheeler Company

on 110-, 220-, or 500-volt circuits and giving speed variations of 6 to 1, 5 to 1, 4 to 1, 3 to 1, 2 to 1, or 1½ to 1. The field yoke is of the best electrical steel cast in one piece. The main poles are made of cast electrical steel or of laminated steel and are bolted to the field yoke. They are skewed along the axis to prevent noise and to

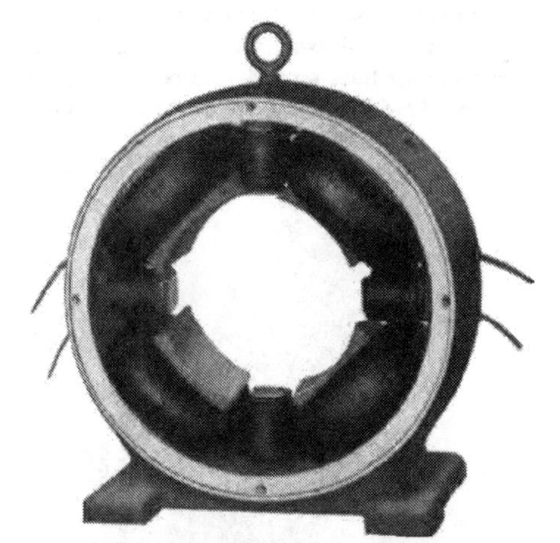

Fig. 86. Electro-Dynamic Motor Showing Interpoles

Fig. 87. Interpole Motor Armature Showing Ball Bearings
Courtesy of Electro-Dynamic Company

permit the gradual approach to maximum field strength under the poles. The interpoles are made of wrought iron or electrical steel. Main and interpole field coils are form-wound. The armature is of the ventilated ironclad drum-wound type, the coils being held in the slots by wedges. These motors are furnished either with ring oiler or ball bearings, as shown in Fig. 87. The end housings are so

Fig. 88. Interpole Motor Applied to Electric Elevator Service
Courtesy of Electro-Dynamic Company

designed that the motor may be semi-enclosed or enclosed. It is also manufactured with vertical shaft. Fig. 88 shows one of the many applications of this machine. These same frames with different field windings furnish a line of constant speed shunt-wound motors.

Fairbanks, Morse & Company. This company manufactures several lines of constant speed shunt-wound motors, using the same frames as for their lines of belted generators. The separate motors range from 150 horsepower at 650 r. p. m., like Fig. 38, down to 2 horsepower at 1800 r. p. m., like Fig. 37. The same frames are also compound-wound, giving a drop in speed of about 20 per cent between no load and full load. They are wound to operate on 115-, 230-, or 550-volt circuits. The smaller sizes are made in all the varieties of open, semi-enclosed, or enclosed, and arranged for floor, ceiling, side wall, or vertical mounting. Fairbanks, Morse & Company also furnishes a line of adjustable speed motors with speed variation of 2 to 1 or 4 to 1, as called for. Their frames, similar to Fig. 37, also furnish a line of commutating pole motors. Fig. 89 shows one of their smaller motors driving an exhaust fan.

Fig. 89. Fairbanks-Morse D. C. Motor Driving Exhaust Fan

Fort Wayne Electric Works. For the largest direct-connected, constant speed motors, such as would be needed for driving air compressors, for instance, this company employs the frames of their engine type generators shown in Fig. 42. Their medium capacity, belted, constant speed motors from 25 to 105 horsepower, shown in Fig. 90, are practically identical with their belted generators. These sizes are 6-pole machines and are made for slow, medium, or moderate speeds operating on 115-, 230-, or 500-volt circuits. Their Northern type B motors, shown in Fig. 91, serve for their smaller sizes. The yoke or field frame is of soft cast steel, circular and in one piece. The poles are of laminated sheet steel through bolted into the frame. The field coils are form-wound and rendered moistureproof. The armature is of the slotted drum type with ventilating ducts. The form-wound armature coils are held in place by

Fig. 90. Fort Wayne 6-Pole Belted Motor

Fig. 91. Fort Wayne Northern Type B Motor

binding bands recessed flush with the armature surface. The commutator is mounted upon and keyed directly to the shaft. The bearings are carried by end bonnets attached to the frame casting by four cap screws. By changing the end flanges through 90 degrees

Fig. 92. Fort Wayne Type K Motor Field Coils and Frame

or 180 degrees, the same motor can be arranged for side wall or ceiling mounting. This line can also be arranged as semi-enclosed and totally enclosed motors, as well as furnishing vertical types.

Fig. 93. Fort Wayne Type K Motor Driving Electric Crane
Courtesy of Toledo Bridge and Crane Company

When furnished with interpoles or, as this company calls them, regulating poles, they become adjustable speed motors.

Fort Wayne Electric Northern type K motors are series-wound and reversible. They are built to operate on 110, 220, or 500 volts and are rated in sizes from ½-horsepower to 100-horsepower capacity on a basis of thirty minutes' continuous operation. They are 4-pole with rectangular cast steel frame of box-like appearance. This is shown in Fig. 92. The bearings are carried by the end plates or bonnets. Fig. 93 shows a crane trolley operated and controlled by these motors.

General Electric Company. Constant speed motors of medium capacity are manufactured by this company, using the frames of their 4- and 6-pole belted generators shown in Fig. 51. The slow speed line furnishes sizes from 250 to 20 horsepower and the moderate

Fig. 94. General Electric Semi-Enclosed D L C Motor

speed from 350 to 30 horsepower. The speeds of the slow speed line are between 925 and 425 r. p. m., while those of the high speed line are from 1250 to 650 r. p. m. The commutating pole design permits the construction of a machine of comparatively large capacity and light weight. The running temperatures are kept low by the free circulation of air through all the parts, the armature and commutator being thoroughly ventilated by means of a fan mounted on the pulley end of the armature shaft. These machines are furnished as open, semi-enclosed, enclosed ventilated, and totally enclosed. A semi-enclosed machine is shown in Fig. 94, and one with forced ventilation is shown in Fig. 95.

For the smaller machines the General Electric Company has a line of commutating pole machines, called type C V C, from 2- to 20-horsepower ratings. A feature of these machines is the field windings of rectangular, cotton covered wire, wound on horn fiber spools. The main field coils are also armor-wound with a single layer of enameled copper wire, serving the double purpose of protecting the active winds from mechanical injury and assisting to a better degree of heat radiation than would be possible with the old style

Fig. 95. D L C Motor Totally Enclosed and Ventilated with Involute Type of End Shield
Courtesy of General Electric Company

taped or cord protected coil. This line allows of any style of mounting or any degree of enclosure. The field windings may be shunt series or compound, giving rise to constant, adjustable, or varying speed motors. The direct-current mill motors brought out by the General Electric Company are of octagonal frame and fireproof construction in five sizes from 30 to 150 horsepower at 230 volts. They employ interpoles and are series-, shunt-, or compound-wound as desired. One of this line is illustrated in Fig. 96.

Fig. 96. General Electric Mill Motor Showing Armatures Being Lifted from Frame

Fig. 97. General Electric Railway Motor

Another complete line of motors is that of the General Electric railway motors, the latest forms of which also use commutating

Fig. 98. General Electric Railway Motor Showing Method of Ventilation

poles and forced ventilation obtained by means of a centrifugal fan integral with the pinion and armature core head. The outer appearance and method of ventilation are shown by Figs. 97 and 98. Besides their larger machines, this company also puts upon the market a complete line of fan motors and small power motors from $\frac{1}{16}$ to $\frac{1}{4}$ horsepower, inclusive. These latter are known as drawn-shell type motors. The frame is punched out of soft steel and then forced into shape. The yoke and pole pieces are made of punched laminations, as shown in Fig. 99.

Fig. 99. Laminated Field of General Electric Drawn-Shell Electric Motor

Holtzer-Cabot Electric Company. The belted generators manufactured by this company, illustrated in Fig. 57, become, with proper windings, their constant speed motors of 1 to 160 horsepower. These frames, when employing a heavy compound winding, become varying speed motors suitable for elevator service in capacities from 5 to 45 horsepower. In this line, however, the base and pedestals are dispensed with and the bearings are held by brackets bolted to the frame. The company's type C machine, shown in Fig. 100, gives them sizes from ⅓ to 30 horsepower to be run at 115-, 230-, or 550-volt circuits. They can be arranged in any position and for any degree of ventilation. They can be built as adjustable speed motors for a speed variation of 1½, 2, 3, or 4 to 1 by field weakening. This

Fig. 100. Holtzer-Cabot D. C. Motor, Open Type

Fig. 101. Reliance Adjustable Speed Motor
Courtesy of Reliance Electric and Engineering Company

company also places a number of motors of very small sizes upon the market.

Reliance Electric and Engineering Company. This concern manufactures an adjustable speed motor, shown in Fig. 101, built in sizes from ½ to 30 horsepower and with speed variations as great as 10 to 1 and as small as desired. It is a simple shunt-wound machine, obtaining its speed variation by altering the reluctance of the magnetic circuit of the machine so as to weaken or strengthen the magnetic field. The principal parts of this machine are shown in Fig. 102, a portion of the illustration being in section. The end of the armature shaft on which the commutator is mounted revolves in a sleeve that slides in the journal bracket. This sleeve is moved back and forth by means of a forked lever controlled through a rod and nut, by the screw on the spindle of the handwheel. The helical

Fig. 102. Part Section of Reliance Adjustable Speed Motor

compression spring surrounding the lever rod always balances the magnetic pull that is exerted by the poles on the armature core. The armature core is slightly tapered, the commutator end being larger in diameter than the other end, and the pole faces are bored to the same taper. When the armature is drawn towards the journal bracket on the commutator end, the air gaps are increased in length and decreased in area. This increases the reluctance, thereby weakening the field and increasing the speed.

Sparking at the brushes when operating at weak fields is prevented by means of special commutating poles midway between the main poles. These commutating poles are in series connection with the armature and laterally displaced from the main poles on the side toward which the armature is withdrawn. The machine is so designed as to give sparkless commutation at all loads at any speed in either direction. The brush rigging is mounted on the end of the sleeve containing the ball bearing, and therefore this bearing, brush rigging, and armature move in unison with no lateral displacement of the brushes on the commutator. The driving end of the shaft

Fig. 103. Stow Multispeed Motor of the Semi-Enclosed Type
Courtesy of Stow Manufacturing Company

slides in a sleeve that revolves with it but does not slide endwise; the driving gear, coupling, or pulley are mounted on the end of this sleeve.

This company also puts a line of constant speed motors on the market ranging from ½ to 50 horsepower; 15 horsepower and smaller are wound for either 115 or 230 volts, while the larger sizes are wound only for 230 volts.

Stow Manufacturing Company. The Stow multispeed motor, as it is called, really belongs to the adjustable speed class, obtaining its speed variations by changing the reluctance of the magnetic circuits. It is bipolar in sizes from ¼ to 4 horsepower, and 4-pole from 4 to 20 horsepower. Fig. 103 shows one of the 4-pole type. The pole

Fig. 104. Sturtevant 4-Pole Motor Driving Ventilating Fan

cores are made hollow and provided with iron or steel plungers, the position of which is adjustable through pinions and worm gears operated by a large handwheel placed on the top or the side of the machine as preferred. When the plungers are withdrawn, the total flux decreases because of the lengthening, and because of the decrease in the effective area of the air gap and also the decrease of effective metal in the field cores. The speed must necessarily increase with

decrease of flux. Also, when the plungers are withdrawn, the flux is along the polar edges, thus maintaining a strong commutation field and also decreasing armature reaction by increasing the reluctance of the armature flux. These effects allow sparkless commutation at any load at any speed within the limits of the design. These motors can be designed to give a speed variation of 3 to 1.

B. F. Sturtevant Company. This concern places two lines of motors on the market. Their 8-pole motors are made in sizes from

Fig. 105. Westinghouse Type S Motor Driving Albro-Clem Electric Elevator
Courtesy of Albro-Clem Elevator Company

9 to 225 horsepower. Those above 40 horsepower are the same in appearance as the generators shown in Fig. 63. The smaller sizes employ three arm-bearing brackets in place of the pedestals. Sturtevant 4-pole motors are built in nine sizes of frames from 1½ to 35 horsepower, and their external appearance is as in Fig. 104. These are all constant speed machines.

Triumph Electric Company. For their larger sizes in constant speed motors this company, like the others, employs their belt-driven generator frames of the design shown in Fig. 67. Six frames are bipolar from ½ to 5 horsepower and eight frames are 4-pole from 7½ to 40 horsepower. They can be wound shunt, series, and compound, thus becoming besides constant speed motors, also varying and adjustable speed motors.

Westinghouse Electric and Manufacturing Company. A line of motors, designated type S, consists of thirteen frames. Used as constant speed motors, they have ratings from 2 to 75 horsepower at 110 volts, from 2 to 150 horsepower at 220 and 500 volts, and from 6 to 100 horsepower at 600 volts. They are mounted in any of

Fig. 106. Westinghouse Interpole Railway Motor with Lower Frame Down for Inspection

the four positions and built open, partially enclosed, or totally enclosed. They can be used also as adjustable speed motors for speed variations of 1 to 1½ or 1 to 2. They are likewise employed as elevator motors, Fig. 105, when compounded by adding a series winding used in starting but cut out for normal running. Adding auxiliary poles, or interpoles, to these shunt-wound machines gives adjustable speed motors of greater speed variation, of ½ to 23 horsepower at a speed ratio of 1 to 4, of ½ to 50 horsepower at a speed ratio of 1 to 2.

This company also manufactures small power motors and a complete line of fan motors. The small power motors are built in three sizes of $\frac{1}{20}$, $\frac{1}{12}$, or $\frac{1}{8}$ horsepower at speeds from 1000 to 2500 r. p. m. and for 110 or 220 volts. They are similar in construction to the bipolar type R machines, fully enclosed, and either series- or shunt-wound. Westinghouse fan motors are of drawn steel construction and series-wound for 110 or 220 volts with speeds from 650 to 2100 r. p. m. Some of their smallest fan motors are wound for 30 volts.

The Westinghouse Company also puts upon the market lines of railway motors, mill motors, and hoisting motors. Their railway motors are built with the usual moistureproof and dustproof cast steel frames. They are 4-pole series-wound for 500 to 750 volts.

Fig. 107. Westinghouse Hoisting Motor with Provision for Back Gear Arrangement

The latest types have the main poles centered at 45 degrees from the horizontal plane and employ interpoles, as shown in Fig. 106. The field windings are of flat copper strap with the turns separated by asbestos ribbon. Westinghouse mill motors are of very similar design, in nine sizes from 5 to 150 horsepower, but wound for 220 volts. They are series- or compound-wound as desired; in the latter case, the shunt winding is so designed as to keep the no-load speed down to about twice full-load speed. This company's hoisting motors consist of 10 frames, wound for 110, 220, or 500 volts, ranging from 2 to 52 horsepower. They are 4-pole, series-wound, full enclosed. Fig. 107 shows a type K direct-current crane motor arranged with back shaft and gear.

DYNAMOTORS, MOTOR-GENERATORS, BOOSTERS

Dynamotor. *Characteristics.* A dynamotor is a transforming device combining both motor and generator action in one and the same magnetic field, with an armature having two separate windings and independent commutators. This class of machines has certain advantages resulting from both generator and motor windings being on the same core. The armature reactions of the two windings, being opposite, neutralize each other, since these windings are on the same armature core. There is, therefore, no shifting of the brushes required and no tendency to spark with varying loads, with resulting ability to stand heavier overloads. They are slightly more efficient than motor-generators, since energy is saved in magnetizing the fields and there is less loss in the bearings, because all torque strain upon the shaft is eliminated. They are also cheaper, of less weight, and more compact. On the other hand, the voltage of the generator can not be varied to any extent except by introducing ohmic resistance into either generator or motor armature circuits. Also the gener-

Fig. 108. Crocker-Wheeler Dynamotor

Fig. 109. Westinghouse Motor-Generator Set

ator voltage can not be maintained absolutely constant as there is no way of correcting for the armature drop.

Uses. Dynamotors are used principally in place of batteries in telegraph main stations, for charging storage batteries in central energy telephone stations, and for electrocautery and electroplating work. This type of machine is made only in the smaller sizes, the output being rarely as high as one kilowatt. The motor winding is arranged for any of the standard voltages, and the generator winding delivers from 6 or 8 to 30 or 40, or else from 100 to 600, volts, depending upon the special use to which the machine is to be put. Fig.

Fig. 110. C & C Welding Set with Switchboard Attachment

108 shows a Crocker-Wheeler dynamotor. A type of dynamotor largely used by telephone companies for ringing telephone bells has the generating winding furnishing current through collector rings and therefore has only one commutator, the one on the motor side.

Motor-Generator. *Characteristics.* A motor-generator is a transforming device consisting of a motor mechanically connected to one or more generators. Being any combination of standard machines, motor-generators come in any capacity. They find their most general application in transforming from alternating current to direct current, and the reverse. That is, one of the machines is an alternating-current one and the other is a direct-current one. It is only in the smaller sizes and in special applications that both machines are of the direct-current type. Fig. 109 shows a 500-kw.

Fig. 111. Motor-Generator Set 500 Volts D. C. to 125 Volts D. C
Courtesy of Crocker-Wheeler Company

motor-generator set, consisting of a 3-wire direct-current generator, driven by a 3-phase synchronous motor.

Use as a Welding Set. A special adaptation of the device is illustrated in Fig. 110, which shows a rather unique C & C welding set with switchboard attached. The outfit is of 300 amperes capacity, and is arranged for double-circuit operation, that is, it supplies two separate welding circuits, the voltage of which may be adjusted independently. It thus permits of having one operator weld with

Fig. 112. View of Disassembled General Electric Balancer set with Single Shaft

metallic electrodes, which require a voltage of only 10 to 30 volts at the arc, while another operator may be using graphite electrodes which require from 40 to 60 volts at the arc. Each circuit is provided with an automatic relay, which, as soon as the circuit is closed, automatically inserts a resistance in series, and thus prevents a

Fig. 113. General Electric Balancer Set

short circuit on the machine. Another set for obtaining 500 volts from a 120-volt circuit, or the reverse, is the Crocker-Wheeler set illustrated in Fig. 111. In the smaller sizes the set is very often arranged with continuous shaft and only two bearings. Fig. 112 shows a General Electric Balancer set disassembled.

Fig. 114. C & C Balancer Set

Fig. 115. Allis-Chalmers Balancer Set

Use as Balancer. Motor-generators having both machines direct-current and of the same voltage find their greatest application in connection with the three-wire system for furnishing the neutral point for connection to the neutral wire, the two outside wires being fed from a single two-wire generator. The two armatures are connected in series between the outside mains, and the common point between the two armatures is connected to the neutral. Thus arranged, the set is also termed a balancer, a balancer set, or a com-

Fig. 116. Crocker-Wheeler Three-Unit Multiple-Voltage Balancer

pensator. The voltage of each machine is equal to that between the neutral and outside wires of the three-wire system. When the load is balanced, both machines operate as unloaded motors, but when unbalanced, one machine operates as a motor, the combined current of the two machines compensating for increased load on the generator side. Balancer sets are generally flat compounded so as to keep the voltage on each side equal, irrespective of the load. An adaptation of this kind of the smaller line of belted generators

Fig. 117. General Electric Shunt-Wound Booster 40 to 65 Volts, Direct-Connected to 250-Volt Motor, Electric Storage Battery Company

brought out by the General Electric Company is shown in Fig. 113. Since the sets are somewhat enclosed in the middle, a fan is provided

Fig. 118. Ridgway Booster Set

and mounted between the armatures. Fig. 114 shows a C & C balancer set composed of their type SL machines. For various systems of multiple voltage motor control, the two armatures may be of different voltages, usually 90 and 160 on a 250-volt main circuit. A three-wire Allis-Chalmers balancer set is shown in Fig. 115. In connection with four-wire systems the set is composed of three machines like the Crocker-Wheeler one shown in Fig. 116, with each machine giving a different voltage, so that six different voltages can be impressed on the armature of the motor run from the four-wire system by selecting the voltages of machines 1, 2, 3, 1 and 2, 2 and 3, and all three.

Booster. A booster is a machine inserted in series in a circuit to change its voltage. It may be driven by an electric motor (in which case it is termed a motor booster), or otherwise. These machines are employed for purposes of voltage regulation in connection with direct-current electric lighting, power, and railway circuits, and with storage battery applications. The voltage of these circuits falls off considerably at points distant from the station with increase of load, and the booster is connected with the circuit in such a way that its voltage is added to that of the circuit, keeping the voltage at the distant points constant. In nearly all cases they are motor-driven, the motor being a shunt-wound machine and the generator a series-, shunt-, or differential-wound type.

Fig. 117 shows a booster of General Electric manufacture composed of a shunt-wound generator and a shunt-wound motor used for storage battery charging and regulation. This requires adjustment to keep the charging of the battery at the proper rate. The whole operation becomes automatic by employing on the generator differential windings properly proportioned so that, up to a certain load, the line voltage is raised by the booster sufficiently to charge the battery, while at higher loads the battery assisted by the booster will discharge into the line.

Fig. 118 shows a booster set manufactured by the Ridgway Dynamo and Engine Company, consisting of a shunt motor and two series generators. Each generator is connected into one of the outside wires of a three-wire system supplying light and power to a distant point.

ENGINE-DRIVEN MULTIPHASE ALTERNATOR WITH DIRECT CONNECTED EXCITER

Courtesy of Fort Wayne Electric Works

TYPES OF GENERATORS AND MOTORS

PART II

ALTERNATING-CURRENT TYPES

General Classification. The class of dynamo-electric machinery employing or furnishing electric energy in the alternating-current form can be subdivided into alternators, motors, rotary converters, and motor generators.

An *alternator* is an alternating-current generator, either single-phase or polyphase. A *converter* is a machine employing mechanical rotation in changing electrical energy from one form to another.

ALTERNATORS

General Characteristics. Alternators are either single-phase, two-phase (also called quarter-phase), or three-phase, depending upon the number of voltages they generate. At the present time, single-phase alternators are no longer being manufactured, the companies furnishing for this purpose a standard three-phase machine. By using any two terminals it can be loaded as a single-phase machine to about 70 per cent of its three-phase rating.

Capacities, Speeds, Number of Poles. The standard sizes in which alternators are manufactured vary from 7½ k. v. a. belted type to 20,000 k. v. a. steam turbine driven. Alternators are rated in kilo volt amperes (abbreviated k. v. a.) instead of kilowatts, since by doing this the question of the power factor of the load is eliminated. For example, a 250 k. v. a. alternator is 250 k. v. a. capacity at any power factor, but 250 k. w. capacity only at unity power factor. Like in direct-current generators, the larger the capacity of the alternator the lower the speed and the larger the number of poles are liable to be. In the case of alternators, however, the speed and the number of poles must be definitely related to each other so as to give a certain frequency. The standard frequencies today are 60 and 25 in this country. For 60 cycles, the number of poles multiplied by the speed in revolutions per minute must equal 7200, while for 25 cycles their product must always equal 3000.

In belted alternators the speeds usually fall between 1800 and 600 r. p. m., while their number of poles are from 4 to 12. Alternators direct driven by steam engines, gas engines, or water wheels run from 900 to 72 r. p. m., and carry from 8 to 72 poles. Turbo-

Fig. 119. Engine-Driven, Revolving-Field, Alternating-Current Generator
Courtesy of Allis-Chalmers Manufacturing Company

alternators, or alternators that are direct driven by steam turbines, run from 3600 to 500 r. p. m., and have from 2 to 12 poles.

Classifications. Alternators may be divided into three types, depending upon the mechanical arrangement of the armature and the field:

(1) Alternators with revolving armatures and stationary fields.

(2) Alternators with stationary armatures and revolving fields.

(3) Alternators in which both armature and field windings are stationary. A revolving part called the inductor causes the flux from the field windings to sweep across the armature conductors.

The revolving-field type is practically the only one manufactured today, the inductor type having been discontinued, and the revolving armature being restricted to the smaller sizes and lower voltages.

DESCRIPTIONS OF TYPES

Allis-Chalmers Manufacturing Company. The alternators put upon the market by this company include five different types:

(1) Standard engine type, in which the rotor is separate from the engine flywheel, as shown in Fig. 119.

(2) Flywheel type, Fig. 120, in which the field poles are mounted directly

Fig. 120. 1500 K. V. A. Alternator Installed for the Mutual Electric Company, San Francisco, California.
Courtesy of Allis-Chalmers Manufacturing Company

on the face of the engine flywheel which then serves the double purpose of flywheel and rotor spider. This is only built in the largest sizes, so that the rotor can give the necessary flywheel effect.

(3) Water-wheel type, having a horizontal shaft, two bearings and flange coupling, as shown in Fig. 121.

(4) Belted type, in sizes from 50 to 900 k. v. a. for use in smaller and industrial plants. This type is illustrated in Fig. 122.

(5) Turboalternators with horizontal shaft, being totally enclosed and employing forced ventilation, as shown in Fig. 123.

All of these types are built revolving field, 60 or 25 cycles, two-phase or three-phase and employ 120 volt field excitation. The

Fig. 121. 5000 K. V. A., 6600 Volt, Three-Phase Water-Wheel Type Alternator Built for Northern California Power Company

Fig. 122. Allis-Chalmers Type AN Belted Two-Bearing Alternator—Stator Moved Sideways

standard voltages for which these machines are designed are 2300, 4400, 6600, or 13200 for power station work and 240, 480, or 600 for local distribution in factories.

Characteristics of Types Nos. 1, 2, 3, and 4. In all but turbo-alternators, the armature or stator yoke is of iron, cast in one piece for the smaller machines and in two or more for the larger ones. The sections of the yoke are bolted together with bolts on the inside of the yoke. To increase the ventilation, the yoke is provided with many cored openings through which the air currents set up by the revolving field can easily pass. Shields are fastened to the sides of

Fig. 123. Allis-Chalmers Turbogenerators Installed for the Pacific Mills

the yoke to protect the armature coils where they project beyond the core. The stator core is built up of steel laminations that are securely clamped between end plates, after being carefully annealed and japanned, so as to reduce the core losses. Ventilating segments are placed at intervals throughout the core so as to provide ducts through which air is forced by the rotation of the field, thereby keeping down the temperature. The form-wound armature coils are interchangeable, each coil or winding unit being completely insulated before being put in place. No insulating material is placed in the slots, the insulation being preferred integral with the coil.

After the coils have been covered with insulating materials and treated with insulating compound, the parts that are to lie in the slots are pressed to exact size in steam-heated moulds. The ends

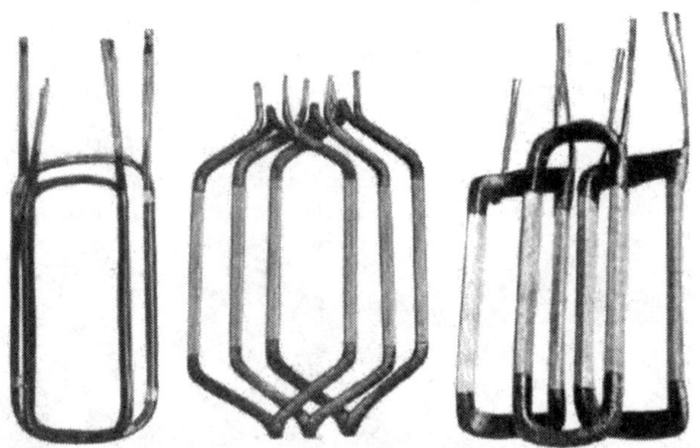

Fig. 124. Form-Wound Interchangeable Armature Coils

of the coils where they project beyond the slots are heavily taped and, when necessary to withstand the stresses due to short circuits, suitable supports are provided. The details of the coils and windings are shown in Figs. 124, 125, and 126.

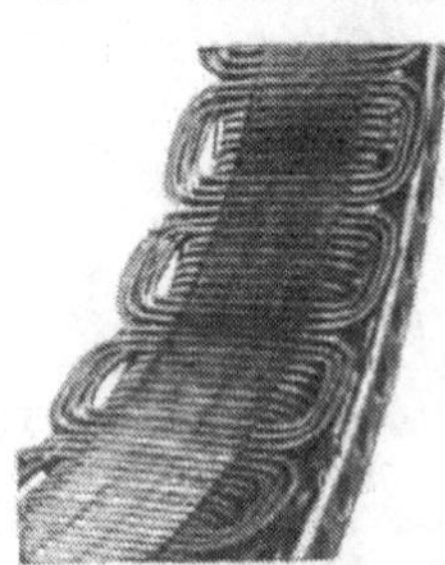

Fig. 125. Allis-Chalmers Chain Winding

In the engine type, water-wheel driven, and belted machines, the field structure consists of laminated pole pieces mounted on the rim of the cast-steel spider; in the flywheel machines the poles are mounted directly on the flywheel face. In some of the large machines the rim of the revolving field is built up of steel laminations. In most machines the pole pieces are provided with dovetail projections that fit into corresponding slots milled in the spider rim and are securely held in place by tapered steel keys. The poles are usually built up of steel punchings clamped together between malleable-iron end plates securely held by rivets.

Each pole piece carries a magnetizing coil held firmly in place between the spider or flywheel rim and the projecting parts of the pole and end plates. The field coils are made of copper strip, edgewise-wound. The collector rings, through which exciting current is supplied to the field, are of cast copper. Current is led into the rings by means of carbon brushes, at least two per ring. Copper shunts or pigtails are attached to the brushes to prevent current from passing through the springs. With engine, flywheel, and water-wheel types the brush holder studs are mounted on a stand, as is shown in Fig. 121. In the belted type the brush holder studs are fastened to the cap of one of the bearing pedestals. The bearings on all machines are of the ring-oil type with ample oil reservoirs.

Fig. 126. Allis-Chalmers 2-Layer Winding

Characteristics of Turboalternators. In the turboalternators, the armature or stator construction is similar to that of the other types. The rotor or revolving field is built up of either steel laminations or of nickel-steel forgings. The slots formed in the core to receive the field coils are radial. The field winding is made of carefully insulated coils of flat strap copper firmly held in the slots by means of bronze wedges. The ventilation of the ends of the coils

Fig. 127. Rotor of a 5625 K. V. A. Allis-Chalmers Turbogenerator

is attained by having them project beyond the ends of the core. At this point they are firmly held by means of nickel-steel rings. For the purpose of obtaining proper ventilation and for muffling the noise produced by the circulation of the air the machines are totally enclosed and the air is taken in at the ends, passing through fans that discharge it over the end connections of the armature coils and through ventilating ducts of the core to the outlets. Fig. 127 shows an assembled revolving field with protecting nickel-steel rings and ventilating fans in place.

Crocker-Wheeler Company. This company manufactures alternators of the engine, coupled, and belted types as well as turbo-alternators. The coupled type of alternator has its own bearings and a shaft which is connected by a coupling to the shaft of an engine, water wheel, or other source of power. All four types are built revolving field, for two-phase or three-phase at 25 or 60 cycles and for the following standard voltages: 240, 480, 600, and 2300. The field excitation is at 125 volts, allowing the standard C.-W. compound-wound direct-current generators to be used as belted, geared, or direct-connected exciters. In their engine, coupled, and belted types, shown in Figs. 128, 129, and 130, the stator consists of a cast-

Fig. 128. 165 K. V. A. Alternating-Current Generator
Courtesy of Crocker-Wheeler Company

iron frame supporting a laminated core of sheet steel, in the slots of which are placed the stator windings.

General Types. The frame is constructed to allow for the proper circulation of air around the stator coils and core, and out through holes in the external surface of the cast-iron frame. This circulation of air over the windings, around the core, and through the core ducts, insures a uniform and low temperature throughout all parts. The laminations have the slots punched on the inner periphery and are securely clamped together with end flanges.

The stator or armature coils are form-wound and thoroughly insulated. They are laid in insulated slots open at the tops which

Fig. 129. Belted-Type Alternating-Current Generator with Direct-Connected Exciter
Courtesy of Crocker-Wheeler Company

are then closed by magnetic bridges that fit between grooves in the sides of the teeth. These bridges serve the double purpose of firmly retaining the coils in the slots and giving the desired magnetic effect of a closed top slot and thereby permit the use of solid field poles and shoes. The details of this slot construction are clearly shown by figures, under the description of the Crocker-Wheeler induction motors. The solid field poles and pole shoes act as dampers and overcome the tendency of machines operating in parallel to get out of step. The projecting ends of the windings are heavily taped and varnished to increase the insulation and give mechanical protection. Additional protection to the ends of the coils is given by light iron grid shields where a separate bearing pedestal construction is used, and by the bearing brackets themselves where that style of construction is used.

For the smaller sizes of alternators, the rotor is a solid steel casting consisting of hub, spokes or web, rim, and poles. Steel pole shoes are bolted to the projecting ends of the poles. For the larger sizes of generators the rotor is like those for the smaller sizes except for the fact that the poles and pole shoes are cast in one piece and bolted to the machined surface of the rotor rim. The rotor or field coils, with the exception of a few sizes that are wound with wire, consist of strips of copper wound on edge. Each turn is properly insulated from the next, and the whole coil is compressed and baked into a compact unit. The coils are well insulated from the neighboring metal and securely held in position by the pole shoes. The exciting current is fed into the field winding through carbon brushes and cast-iron collector rings. These rings are supported on a cast-iron hub from which they are suitably insulated. In the belted type, however, one collector ring is placed on each side of the rotor. The brush holders are of the radial type, with adjustable tension, and the brushes are self-feeding. The entire brush rigging and yoke

Fig. 130. 2000 K. V. A. Coupled-Type Alternator
Courtesy of Crocker-Wheeler Company

are mounted on the bearing or bearing bracket. All bearings are ring-oiling, with caps that are easily taken off to permit the removal of the journal boxes. In the larger sizes the stator frame and bearing pedestals are bolted to the cast-iron base. In the smaller sizes the bearings are carried by three arm brackets bolted fast to the ends of the stator housing.

Turboalternators. Crocker-Wheeler turboalternators range from 500 to 6000 k. v. a. They have a frame forming the housing for

Fig. 131. 3750 K. V. A. 2200 Volt, Three-Phase Crocker-Wheeler Alternator

the stator punched laminations that is ribbed and braced on the inside to obtain strength and rigidity. Both ends of the stator housing are closed by means of shields bolted fast to it. The shields serve as passage ways to carry the incoming ventilating air where it is needed, protect the windings of the stator from injury, and prevent the admission of oil, dust, or dirt. These enclosing shields also reduce the noise made by the rapidly-revolving rotor.

The method of ventilation in these machines provides for drawing the air in through horizontal openings on the under side of each end shield and discharging it through openings in the lower central part of the frame. The absence of other openings and of any projecting lugs or ribs results in a frame of smooth exterior surface and of compact and symmetrical shape as shown in Fig. 131.

The laminations of the stator are of sheet steel securely clamped together by end flanges and provided with ventilating ducts to secure uniform and low temperature throughout the core. The insulated,

Fig. 132. Rotor of 250 K. V. A. Two-Phase Crocker-Wheeler Turboalternator

form-wound stator coils are held in the open slots of the stator by wooden wedges fitted into the grooves in the teeth. The projecting ends of the windings are heavily insulated and varnished and are further protected by the extended portion of the stator housing and the end shields.

In some cases the rotor core is a solid steel casting with pro-

Fig. 133. Rotor of 3500 K. V. A. Three-Phase Crocker-Wheeler Turboalternator

jecting poles to which the pole shoes are bolted, as shown in Fig. 132. These pole shoes provide proper distribution of the magnetic flux and also hold the field coils in position on the poles. The coils are wound with edgewise copper, insulated between turns and also from the pole and the shoes. In another construction shown in Fig. 133, the rotor core is built up of steel punchings or disks keyed to the shaft so as to prevent shifting endwise. The rotor windings consist of conical coils of copper strip, wound on edge and laid in radial slots. The coils are insulated by mica and fibrous material between turns as well as from the rotor body. Non-magnetic

Fig. 134. 150 K. V. A. Engine-Type, Revolving Field, Alternator, Direct Connected to Steam Engine
Courtesy of Electric Machinery Company

metallic wedges, fitting in deep grooves in the sides of the slots, retain the coils in place. Hollow cylinders fitted at the ends of the core hold the coils firmly in position. The ventilating air is forced through bronze rings at the ends of the cylinders. The design of the rotor provides for drawing air in from both ends along the shaft toward the interior, whence it is discharged through ventilating spaces. It then passes through the stator ducts and stator coils into the stator housing, where it is guided around the core through an ample space and discharged below the frame. The collector rings for the exciting current are insulated from the hub and so arranged that there is free circulation of air around them.

Electric Machinery Company. The lines of alternators put upon the market by this company are of the engine, coupled, and

Fig. 135. Vertical Type of Electric Machinery Company Alternator

Fig. 136. 100 K. V. A. Two-Bearing Pedestal-Type Alternator
Courtesy of Electric Machinery Company

belted types. They are all revolving-field, two-phase or three-phase, 25 or 60 cycles, employ 125 volt exciters and are wound for 240, 480, 600, 1200, or 2400 volts. One of their engine-type alternators is shown in Fig. 134. Their coupled machines are of two types, horizontal shaft, and vertical shaft, for direct coupling to vertical water-wheel shafts. This latter style is shown in Fig. 135, while Fig. 136 is an illustration of one of their belted types. Their belted machines in the larger sizes are three-bearing and in the smaller sizes have the bearings supported by end brackets instead of separate pedestals.

General Characteristics. Except in the largest sizes, where it is

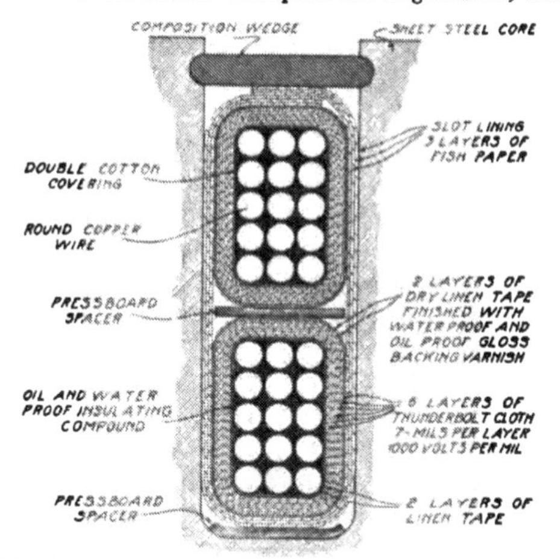

Fig. 137. Details of Electric Machinery Company Armature Coil and Slot Insulation

split horizontally, the circular cast-iron armature ring is one piece. The sheet-steel armature core is built up of laminations assembled with staggered joints and clamped between rigid cast-iron retaining rings. Ventilating spaces are provided at short intervals by means of box-shaped spacers near the ends of the teeth and through the depth of the core. Open-type armature slots carry the form-wound interchangeable coils rigidly held in place by composition wedges, as shown in Fig. 137. At both ends of the armature core, tooth supports reinforce the teeth and prevent humming and chafing of the coils. These are punched steel strips, V-shaped, and riveted to each other at the open ends so as to form a continuous chain of loops, each one supporting one tooth. The revolving field consists of a spider very much like a thick-rimmed pulley. It is made of cast iron for slow speeds but of cast steel for high speeds. This spider carries the pole pieces with their field windings. Square holes are

Fig. 138. Box-Frame Type of Armature
Courtesy of General Electric Company

formed in the rim for receiving the anchors of the pole pieces. The pole pieces are built up of steel laminations held between end plates. The field coils are edgewise-wound copper ribbon. Cast-iron collector rings and carbon brushes allow the exciting current to pass through the field windings.

General Electric Company. *General Characteristics.* The alternators manufactured by this company include complete lines of engine, water-wheel driven, belted, and turbo machines, designed for standard frequencies and voltages. In their engine and water-wheel

Fig. 139. 450 KW. Three-Phase, Engine-Driven Alternator
Courtesy of General Electric Company

driven types, the stationary armature consists of a circular cast-iron frame, supporting a laminated sheet-iron core in which the armature windings are embedded. The frame is either of the box type, Fig. 138, or of the skeleton type, Fig. 139. The laminations are stacked together and held rigidly in place by heavy steel clamping fingers, ducts being provided in the stacking at frequent intervals to allow for the free circulation of air. The outer circumference of the

Fig. 140. Section of G. E. Alternator Showing Method of Dovetailing Core Laminations to Stator Frame

Fig. 141. Section of G. E. Stator Sh.wing Ventilating Ducts

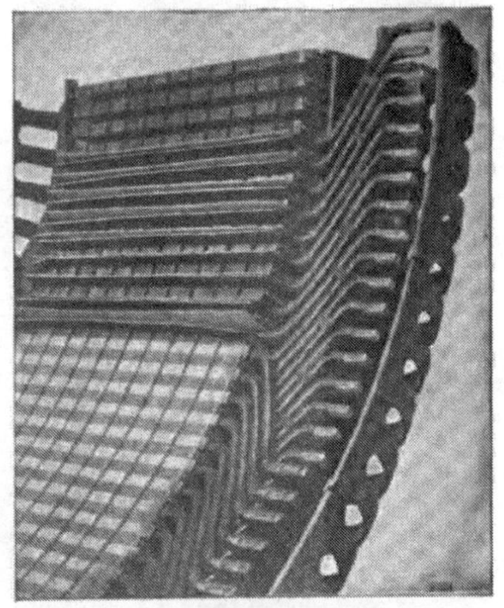

Fig. 142. Section of G. E. Stator Showing Air Ducts and Supporting Fingers Along Slot Projections

laminations is dovetailed for fastening to the frame and the inner circumference is slotted to receive the windings (see Figs. 140, 141, 142, and 143). The armature windings consist of carefully insulated form-wound coils held in open slots by suitable wedges. The coils and windings are clearly shown in Figs. 141, 142, and 143. The revolving-field structure, Fig. 144, consists of laminated pole pieces bolted to a cast-steel or iron ring, which is connected to the hub by arms of ample section, as shown in Fig. 145. The pole pieces, Fig. 146, are built up of laminated iron sheets, spreading at the pole face so as to secure

Fig. 143. Section of Stator Showing Method of Assembling Coils

Fig. 144. Revolving Field with Wire-Wound Coils
Courtesy of General Electric Company

not only a wide polar arc for the proper distribution of the magnetic flux, but also to hold the field windings in place. The laminations are either riveted or bolted together and reinforced by

Fig. 145. General Electric Rotor Spider

two stiff end plates. They are either bolted to the spider or solidly mounted by means of dovetail slots in the rim, Fig. 145, the steel wedges being guarded by two bolted end rings. In the smaller

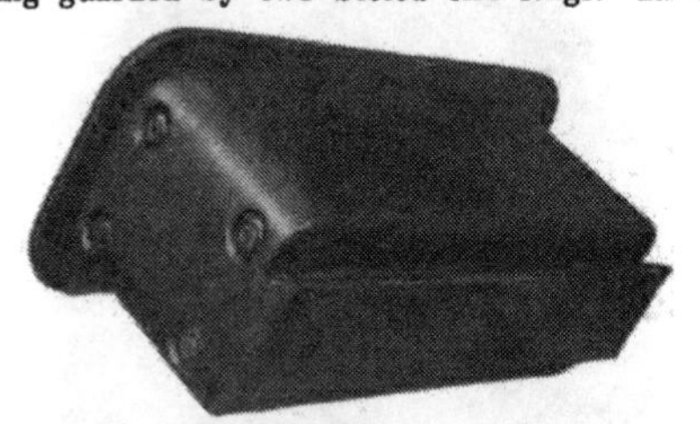

Fig. 146. General Electric Pole Piece Showing Dovetailing

machines the wire is wound on spools which are slipped over the pole piece and held in place by the large tips. The field coils on the larger machines consist of a single strip of flat copper, wound on edge so that every turn has a surface exposed to the air for cooling. The collector rings for the low potential field current are made of cast

RAINBOW POWER STATION, GREAT FALLS, MONTANA
Courtesy of Montana Power Company

iron, are provided with duplicate carbon brushes, and require practically no attention in operation as they are so designed that all surfaces of the rings have easy access to the air, thus insuring good ventilation (see Fig. 147). The brush holders are supported by a cast-iron standard, or yoke, from which they are insulated by suitable bushings.

Fig. 147. General Electric Collector Rings

Gas-Engine-Driven Alternators. For alternators driven by gas engines, the G. E. Company adds a short-circuited squirrel-cage or

Fig. 148. Split-Field Spider with Squirrel-Cage Winding
Courtesy of General Electric Company

amortisseur winding over the revolving field. This is done to decrease variations in angular velocity and improve parallel operation. Any tendency to pulsation or hunting between the engines that is accompanied by a sudden change in the angular velocity of the field, generates current in this short-circuited winding that resists the forces causing pulsation. The appearance of this short-circuited winding is shown in Fig. 148.

Engine Types. The engine-type machines are made in standard sizes from 50 to 2000 k. v. a. The water-wheel driven, arranged for either horizontal or vertical shaft, reach as high as 20,000 k. v. a. and a voltage of 13,200.

Belted Alternators. The standard belted machines manufactured by the G. E. Company are made in two lines. One is in 7 sizes from 7½ to 200 k. v. a. wound for 240, 480, 600, or 2300 volts, two-phase or three-phase. They are built revolving field and the usual methods of construction are employed. The second line differs radically in that the machines are of the revolving-armature type. They are built in three sizes, 7½, 15, and 25 k. v. a. rating wound for 120, 240, 480, or 600 volts, two-phase or three-phase, at 60 cycles. Their general appearance is shown by Fig. 149. The armature contains two distinct windings placed on the same core. The main generator armature winding is connected to the collector rings furnishing two-phase or three-phase alternating currents, while the other winding connected to the commutator furnishes direct current for the fields. These machines, therefore, require no separate or external exciter. The field structure consists of four laminated pole pieces cast into the yoke or frame.

Fig. 149. General Electric 25 K. V. A. Alternator

Turboalternators. The turboalternators of this company are of the enclosed type and self-ventilated. In the larger sizes, a fan on each end of the rotating field draws in air through the ducts at either end of the generator and directly under the end shields which

Fig. 150. Portion of Stator Showing Armature Coils Assembled in Slots for G. E. Turboalternator

Fig. 151. G. E. Turboalternator Showing Armature Coils, Air Ducts, and Laminations

act as funnels. This air is forced through all parts of the generator, cooling the coils, and is then discharged directly downward through

Fig. 152. Revolving Field of G. E. Turboalternator with Coils in Process of Assembly.

Fig. 153. Revolving Field of G. E. Turboalternator Showing Details of Construction

a large central duct. The collector rings and brushes are placed at the end of the generator where they are readily accessible. Fig. 150 shows a view of the stationary armature with part of the coils assembled in the slots. The coils are form-wound, interchangeable, and heavily insulated. They are inserted in the open slots and held in place by wedges dovetailed into the iron punchings. Ample ducts provide free passage to ventilating air. Fig. 151 shows the completely assembled coils and wedges and the rigidly supported end windings. Fig. 152 shows a revolving field with coils in process of assembly. These coils are readily dropped into the slots as shown. They are insulated to withstand high temperatures. The completed field is a compact rotating element of great mechanical strength. As shown in Fig. 153, the surface is practically smooth. The end windings, which necessarily extend beyond the punchings, are held against centrifugal strain by heavy steel retaining bands.

Ridgway Dynamo and Engine Company. The Ridgway Company manufactures complete lines of engine, water-wheel, and belted-

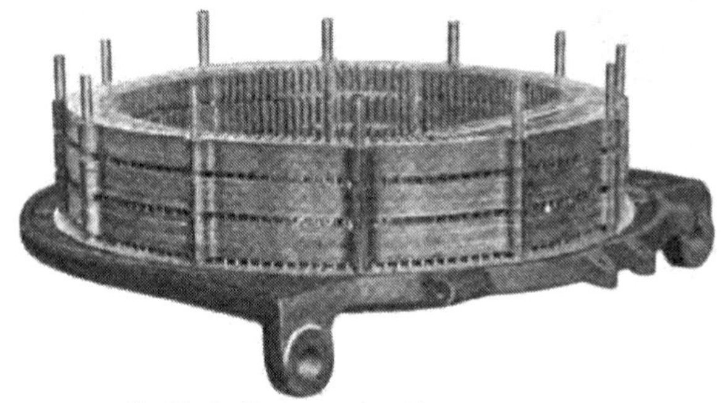

Fig. 154. Stacking Armature Core of Ridgway Alternator

type alternators in sizes 30 to 300 k.v.a. for the belted, and 35 to 750 k. v. a. for the others. They are wound for single-phase, two-phase, or three-phase circuits for either 25 or 60 cycles and for any of the standard voltages.

General Characteristics. The armature core is built from punchings of the highest grade transformer steel. Before being stacked, as shown in Fig. 154, the punchings are machine-coated with insulating varnish for the purpose of reducing eddy-current losses. Several air ducts are provided in the core for ventilating purposes.

The armature frame consists of two heavy cast-iron rings having an I beam section. The core is clamped securely in position between these rings by bolts that pass through the core but outside of the magnetic circuit. This method of construction, clearly illustrated in Figs. 155 and 156, has the advantage of securing splendid ventilation for the core. Neat and substantial guards bolted to the frame, and shown in Figs. 156 and 157, protect both ends of the windings from mechanical injury.

For 1100 volts and over, the winding is of formed coils, impregnated with insulating varnish, and baked into a solid mass. All such coils are wound on a single form and are, therefore, interchangeable. For lower voltages the winding consists of solid copper bars, carefully insulated with tape and varnish and baked before being placed in the slots. The use of bars avoids the necessity of connecting a number of circuits in parallel. Before the winding, whether of bars or formed coils, is placed in the core, the slots are lined with heavy insulating material of high, dielectric strength.

Fig. 155. Ridgway Armature Core Completed

The field spider of small engine-type generators is a single steel casting. In larger machines it is of cast iron, onto which is shrunk a heavy rim of cast steel. The pole pieces are built up of laminated steel punchings, held between heavy brass end pieces and are bolted to the spider by stud bolts which screw into the poles and pass through the rim of the spider. The field of a belted machine consists of a laminated steel core mounted on a heavy cast-iron hub and clamped between substantial end plates. In the periphery of this core are punched T-shaped slots, corresponding to similarly shaped projections on the laminated pole pieces. A small space is allowed between the projections on the pole pieces, and two sides of the

Fig. 156. Ridgway Water-Wheel-Type Alternator

Fig. 157. Engine-Type Alternator Connected to Center-Crank Engine
Courtesy of Ridgway Dynamo and Engine Company

slots, and into these spaces are driven square keys that hold the pole pieces securely in place.

The field coils of small engine-type and belted-type machines are wound with square copper wire, having rounded edges and insulated with a double layer of cotton. On larger machines of both types the field coils consist of copper strip wound on edge, and insulated with paper and insulating varnish. At one side of each coil, there is placed a thin metal blade which, when the rotor is in motion, produces a strong current of air past the coil and across the face of the armature coil. In addition to this fan, there is, in

Fig. 158. 250 K. V. A., Three-Phase Triumph Alternator

belted generators, a ventilating duct passing through the middle of the coil. The collector rings for conducting the exciting current to the field coils are of cast iron, mounted on a separate spider and well insulated therefrom. The brush holders are of the box type, each carrying two carbon brushes, and are usually mounted on the adjacent bearing.

Triumph Electric Company. Triumph alternators may be of the engine, coupled, or belted type, are wound single, two-phase or three-phase for 25 or 60 cycles and for the standard voltages of 240,

480, 600, or 2400. They are all of the revolving-field type with laminated poles carrying edgewise-wound strip field coils. The stationary armatures are made of slotted laminations carrying form-wound interchangeable coils and have the usual ventilating ducts. The field excitation is at 125 volts by means of cast-iron collector rings and carbon brushes. Figs. 158 and 159 show the appearance of these machines.

Westinghouse Electric and Manufacturing Company. *Engine Types.* The standard ratings of Westinghouse engine-type alter-

Fig. 159. Belted Type of Alternator
Courtesy of Triumph Electric Company

nators, of type E, Fig. 160, include a large number of machines ranging in capacity from 50 to 1100 k. v. a., fully covering all synchronous speeds used in ordinary engine practice and carrying either two-phase or three-phase windings.

The stator frame is of cast iron. The box section employed is very rigid, provides the necessary space for the end connections, and also permits excellent ventilation. Transverse ribs on the inner circumference strengthen the frame and form free air passages around the core. The smaller frames are cast in one piece, the larger ones being split horizontally. Feet with ample bearing surfaces are cast upon the frame; shoes and slide rails permit adjustment of position as shown in Fig. 161. The armature core is built up of punchings or laminations of thin sheet steel of high permeability, thoroughly annealed and japanned. These laminations are assembled under pressure in dovetail slots in the interior transverse ribs and securely held in place by finger plates and end plates.

Fig. 160. Westinghouse Type E, 75 K. V. A. Generator Showing Collector End

Ventilating finger plates of sheet steel are assembled with the laminations to form suitable air ducts. Finger plates of malleable iron are used at each end of the core for supporting the teeth. These plates are of greater depth than would be necessary for strength alone, in order to provide generous ventilating ducts along the outer sides of the core between the laminations and the end plates. End plates of cast iron, assembled under heavy pressure outside the finger plates, complete the assembly of the core as a rigid unit, free from all possibility of internal vibrations. In the smaller machines these end plates are keyed to the frames. In the other frames, in which the end plates are segmental, they are held in place by through bolts between the ends, clear of active iron. The armature slots are open and the armature coils are held in place by hard fibre wedges firmly secured in slots of the teeth, effectually preventing vibration. The armature coils are machine-wound of double cotton-covered copper wire or strap. After winding and forming, the coil is dried out in vacuum and filled with an insulating compound under pressure. An outer insulation is then applied, consisting of layers of flexible sheet mica for that portion of the coil within the slots and of treated tape for the ends. The entire coil is then given a further protection of cotton tape and finally treated a number of times with an insulating varnish that protects the insulation and gives a finished appearance. The end bells attached to the frame are sheet-steel segments built up into circular form. They are of light weight and open construction yet rigid and practically indestructible, full protection being afforded by them to the end connections without in any way interfering with perfect ventilation.

Fig. 161. Westinghouse Shoe and Slide Rail for Type E Generator

The brush holders are of the standard sliding shunt type and are supported by cast-iron brackets. On the smaller sizes the brackets are bolted to the armature frame (see Fig. 160). On the larger sizes the brush holders are carried on one or two separate bracket stands, as shown in Fig. 162, supported on the foundation or engine-

Fig. 162. Westinghouse Brush Holder Bracket Stand for Large Size Alternators

bearing pedestal. The spider of the type E alternator rotor is a single steel casting, carefully proportioned with reference to cooling strains. As the rim of the spider forms a part of the magnetic circuit, better magnetic conditions are obtained at the joint between pole and rim with cast steel than would be the case if cast iron were employed. The pole pieces are built up of sheet-steel punchings riveted together and bolted to the spider rim. As an additional means of creating air currents and regulating the temperature of the alternator, small radial steel plates with surfaces at right angles to the direction of rotation are bolted at intervals on each side of the spider rim. Edgewise-wound strap coils are used for the field coils. This construction is preferred, every turn being exposed to the air

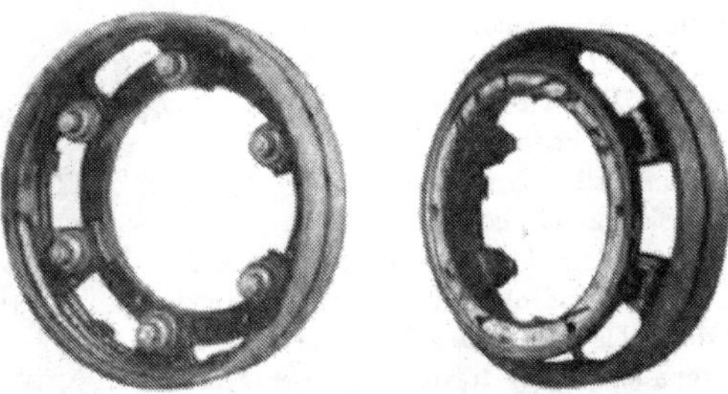

Fig. 163. Front and Rear Views of Type E Collector Rings
Courtesy Westinghouse Electric & Manufacturing Company

so that the heat of the coils is readily conducted to the surface and so dissipated. Between adjacent surfaces of the turns of the strap, layers of flexible fireproof insulating material are inserted and the entire coil is treated with an insulating varnish, making the coils practically fireproof and indestructible.

The collectors, Fig. 163, are of the spider type, consisting of two accurately machined cast-iron rings, mounted on a cast-iron hub from which they are insulated by V-shaped moulded mica bushings, and micarta bushings and washers on the supporting bolts. The assembled collector is bolted to the hub of the spider. Two brushes, at least, are provided for each ring, to deliver current at 125 volts to the field winding. With internal combustion engines as prime movers, a cage-damper winding is provided on the field poles.

This winding, shown in Fig. 164, consists of a series of copper bars embedded in the pole faces with ends short-circuited similarly to the squirrel-cage winding of certain types of induction motors. It serves as an effective damper, tending to prevent hunting when alternators are operated in parallel.

Water-Wheel-Driven Alternators. The water-wheel-driven alternators of the Westinghouse Company embody the same features

Fig. 164. Rotor of Type E Alternator with Cage-Damper Winding
Courtesy of Westinghouse Electric & Manufacturing Company

as type E alternators but differ principally in the construction of the rotors, depending upon the centrifugal stresses produced.

At the lowest peripheral speeds, the rotor construction employed is the same as for the engine-type alternators with bolted poles. Fig. 165 shows a construction wherein poles are dovetailed into the rim of a cast spider. Machines of a larger diameter have their rotors built as shown in Fig. 166, where a laminated rim is dovetailed to a cast spider and the poles are dovetailed to the rim. For the very highest speeds the construction shown in Fig. 167 is employed.

Fig. 165. Westinghouse Rotor Construction Showing Cast Spider and Dovetailed Pole Pieces

Fig. 166. Westinghouse Rotor Construction Showing Cast Spider, Laminated Rim, and Dovetailed Slots

Rolled-steel plates form the spider and the poles are dovetailed to the spider.

The stator and armature constructions of these machines are similar to the engine type. The water-wheel alternators are designed with either horizontal or vertical shafts.

Belted Types. The line of belted machines cover sizes from 30 to 200 k. v. a. at 60 cycles from 240 to 2400 volts, two-phase or three-phase. The three smaller sizes are provided with bracket-bearing housings, while the other sizes have pedestal bearings. In these machines the rotors consist of a laminated spider, as shown in

Fig. 168, built up of thin steel plates assembled upon a mandrel and firmly riveted together under hydraulic power. The poles are built

Fig. 167. Westinghouse Rotor Construction Showing Rolled-Steel Plate Spider and Dovetailed Pole Pieces in Position

up of steel laminations of the same thickness as those of the spider and riveted together. Each pole is dovetailed into the spider and retained by two taper steel keys. The field coils are wound with wire. The stator and armature constructions are practically the same as in the other lines.

Fig. 168. Westinghouse Laminated Spider with Pole Pieces Dovetailed to It

Turboalternators. Westinghouse turboalternators are also of the revolving-field type. This construction avoids all moving contacts between the generator and the main circuit to which it is connected and is of especial advantage in dealing with large currents and high voltages. These generators can be wound to supply single-phase, two-phase, or three-phase circuits of any commercial frequency or voltage. Standard generators are wound to supply two-phase or three-phase circuits of 240, 480, 1200, 2400, 6600, 11000, or 13200 volts at 25 or 60 cycles in capacities up to 15000 k. v. a. They are designed for separate excitation at a standard voltage of 125. Except in the largest sizes the frames are made in one piece. In the large machines the frames are divided horizontally, the two parts having faced joints and being bolted and keyed together so that they form practically a single piece.

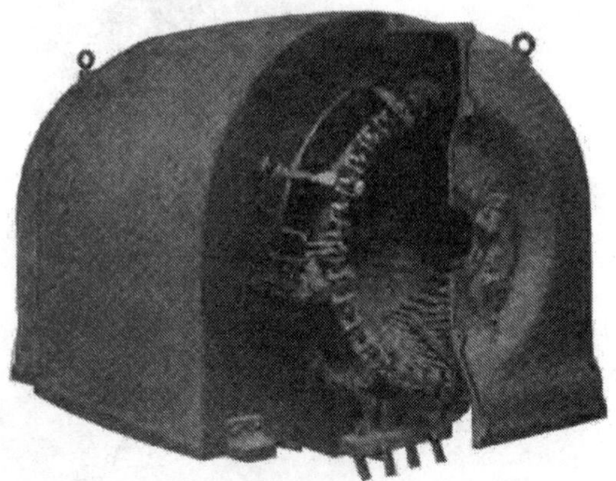

Fig. 169. Turboalternator with Rotating Field Dismounted and Half of End Bell Removed
Courtesy of Westinghouse Electric & Manufacturing Company

The armature or stator is built up of punchings of soft sheet steel. Ventilating plates are provided at suitable intervals, forming air ducts in the core. The core is slotted to receive the armature windings, the shape of the slot depending upon the capacity of the machine and the character of the windings. Either open or partly closed slots are used, the edges of the former being grooved at the top to receive the retaining wedges holding the coils in place. At the ends, the teeth are supported by finger plates and heavy iron retaining plates that are pressed into place and keyed. Form-wound armature coils are used and the winding is of wire, strap, or bar

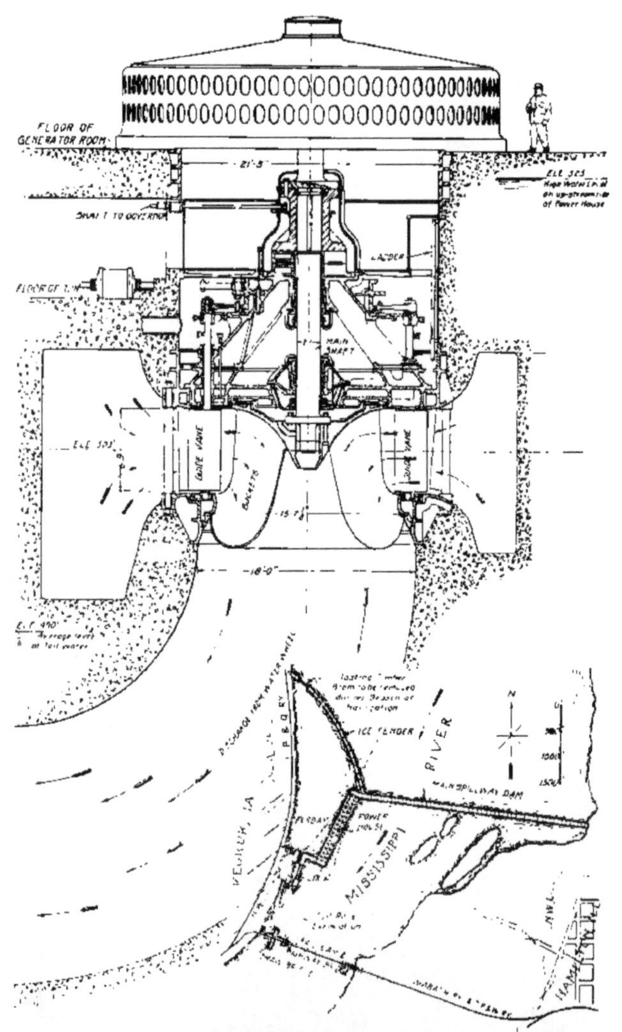

SECTION OF ONE OF THE TURBINES INSTALLED IN THE MISSISSIPPI POWER COMPANY'S PLANT AT KEOKUK

Thirty such turbines, each furnishing over 10,000 horsepower, represent the capacity of the plant when completed. The lower diagram shows plan of the entire development

Courtesy of Mississippi River Power Company

copper, depending upon the capacity and voltage of the machine. The slots are provided with a lining of fibrous material, and the coils are wedged into the slots by wedges fitted in grooves in the sides of the slots or below the overhanging tips of the teeth. The armature coils are firmly braced at the ends of the frame in such a manner as to insure them against displacement, as shown in Fig. 169. Closed end bells are provided which cover the ends of the armature coils and the moving parts of the machine between the frame and the

Fig. 170. Westinghouse Turboalternator Fully Enclosed

bearings with the exception of the collector rings which are external. These end bells further serve to protect the windings of the machine from mechanical injury. The enclosed frame, as seen in Fig. 170, with its end bells provides for a positive ventilating system. The end bells close each end of the alternator and form a duct through which cool air is drawn into the machine and forced out through ventilating ducts in the stator into the interior of the frame, from which the air passes down through the bed plate and escapes. In the large generators the air also escapes through openings in the top of the frame.

The rotor of the machine is provided with a fan at each end to draw the air into the machine. The revolving fields or rotors are constructed with two, four, or six poles, according to the frequency and capacity of the machine. The fields are of small diameter and are designed with special care to avoid windage losses and to facilitate ventilation. The poles of the two-and four-pole rotors, Fig. 171, are machined from disks forming the central body, and the slots to receive the field coils and the grooves for the binding wedges are milled. The six-pole rotors are built up by bolting poles to a central body. The rotors are carefully balanced after they are wound. In some designs, the rotor is pressed and keyed onto the shaft; in others, the shaft is formed of steel, cast or forged integral with the rotor core. For two-pole machines the rotor is generally made from a solid cylinder and the shaft is made in two portions and secured to each end of the rotor by heavy bronze flanges and suitable bolts.

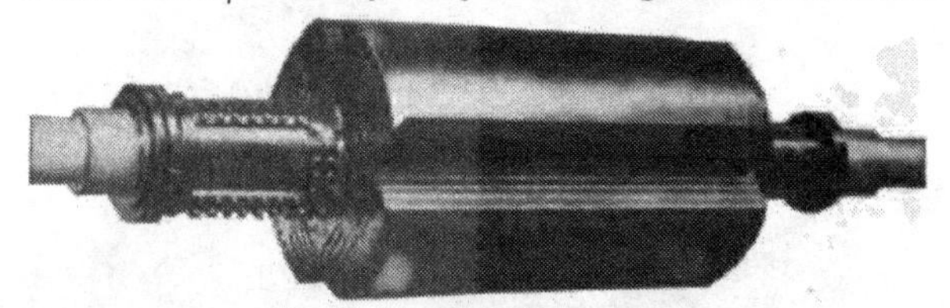

Fig. 171. Westinghouse Four-Pole Rotor

The field winding consists of copper strap embedded in slots cut in the poles. The coils are wound directly in place under a heavy tension. A groove is cut in each side of the slots and brass wedges are driven in to hold the coils in place. The coils are heavily insulated with material of high dielectric and mechanical strength. This material is applied in several layers. The winding and insulation are tightly wedged in place. The exciting current is delivered to the field or rotor winding by means of a pair of collector rings and suitable brushes. The collector rings are made in one piece, with no joints, and are shrunk on the shaft. Either carbon or copper brushes are used.

MOTORS

Classification. The various alternating-current motors used on commercial circuits belong to one of the following classes:

(1) *Synchronous.* This type is really nothing more than an inverted alternator, it being possible to use identically the same

machine for either purpose. As a motor, however, it is practically not self-starting and also must be separately excited from some source of direct-current supply. It may be, of course, either single-phase, two-phase or three-phase.

(2) *Polyphase Induction.* This type has its field windings fed from the alternating-current supply which then produces a rotating magnetic field that makes the machine self-starting. It has no commutator and its armature circuit is not connected to the supply circuit. It may be either two-phase or three-phase.

(3) *Single-Phase Induction.* This type has its field windings but not its armature connected to the supply circuit. It has no commutator and it is not self-starting. It can be brought up to speed by one of the following methods: Hand starting, shading-coil or creeping-field starting, split-phase starting, or repulsion-motor starting.

(4) *Repulsion.* This type is single-phase, with its field windings connected to the supply circuit. It has its armature, commutator, and brushes connected to a local circuit, and whatever current flows in this circuit is entirely induced therein. It has practically the same characteristics as the direct-current series motor, and needs a load to keep it from attaining a dangerous speed. It is not used commercially to any extent although the principles underlying its action are employed in getting many single-phase motors up to speed.

When a second set of brushes and an auxiliary field winding are added, the machine becomes a compensated repulsion motor and has characteristics similar to a compound-wound direct-current motor. In this form it is used for small power motors.

(5) *Single-Phase Series.* This is a machine having its armature and commutator connected in series with its field windings. It has all of the characteristics of the direct-current series motor, and is applicable to the same classes of work. This type of motor is not in general use.

DESCRIPTION OF TYPES

Allis-Chalmers Manufacturing Company. *Induction Motors.* The induction motors of the Allis-Chalmers Company are built in many sizes, from 1 to 300 h. p., wound for two-phase or three-phase, 25 or 60 cycles, and wound for the standard voltages of 110, 220, 440, 550, and 2200. They are divided into two classes: Type AN, or constant speed type, illustrated in Fig. 172, which has a squirrel-cage secondary winding; and type ANY, or the variable-speed type, which has a phase-wound secondary. The stators of the two types of motors are identical. The frame of these motors is of the box type, with large cored openings that permit of the passage of a large amount of air and make the motors cool running. Lugs are cast on the interior surface of the frame to support the stator core, leaving a large air space between the two. Supporting feet are cast in one piece with the frame. The stator core is built up of high grade electrical steel punchings that have been carefully

Fig. 172. Exploded View of Small AN Allis-Chalmers Motor

annealed and coated with insulating varnish. In the large motors these punchings are assembled in the frame with spacers introduced at intervals to form ventilating ducts through which currents of air are forced by the rotating element, thereby carrying off the heat from the interior of the machine. The inner periphery of the punchings is slotted to receive the coils.

The stator coils composing the field windings are form-wound and interchangeable. They are carefully insulated with the best obtainable insulating materials and, after they are wound, are impregnated with an insulating and waterproof compound. The straight portions of the coils are pressed to gauge in moulds so that they fit the slots exactly. Except in some of the smaller sizes below 5 h.p., the coils are placed in open slots. The ends of the coils, where they project beyond the core, are well protected by the stator yoke and end housings.

The rotor core is built up of steel laminations in a manner similar to the stator core. In all but the smallest sizes the laminations or punchings are mounted on a cast-iron spider having arms shaped to act as fan blades for forcing air through the motor. The spider is pressed onto the shaft. In the smallest sizes the punchings are mounted directly on the shaft, which is properly machined to hold them firmly. The rotor A N winding consists of a series of copper bars, each placed in its individual slot and held in place by the overhanging tips of the teeth.

In the smaller sizes round copper bars are used and are fastened to a copper disk, which forms an end ring having holes in it to receive the ends of the bars. The ends of the bars are turned down somewhat smaller than the body and exactly fit the holes in the end ring. The square shoulder formed on the bar fits firmly against the disk. The bars are expanded to completely fill the holes in the disk and the end of each is spun over to form a head. In order to prevent any possible deterioration due to corrosion of the copper, the joint is thoroughly tinned with solder having a high melting point. On motors of large size, end rings and conductors are both of rectangular cross section. The bars are fastened to the end rings by machine-steel cap screws. On the smaller sizes the ring-oiled bearings are made solid, and are carried in the ends of housings or brackets, which are so arranged that they can be rotated 90 or 180 degrees, thus allowing the motor to be mounted on the floor, ceiling, or side wall. In the larger sizes the motors are supplied with separate pedestal bearings and, when necessary, are furnished with an outboard bearing and an extended shaft.

Fig. 173. Allis-Chalmers Rotor of Type ANY Induction Motor

Type A N motors above 5 h. p. are provided with potential starters for reducing the voltage applied to the stator windings during the starting period. The starter consists of potential transformers having several taps each and a controller for making the necessary changes in the connections.

The A N Y motors use the same stators as the A N. The rotor,

Fig. 174. Single-Phase Induction Motor
Courtesy of Bell Electric Motor Company

however, is wound for three-phases and the terminals brought out to three slip rings, as shown in Fig. 173. The front bearing bracket is slightly modified to make room for these rings on the inside. High grade, low resistance brushes, accurately fitted to the rings, are used. This type of motor is, therefore, controlled by inserting in the secondary circuit an external resistance adjusted by means of a suitable controller connected to the slip rings. It may be designed for starting duty only, or for both starting and speed regulation. This controller is non-reversing and is designed to control the secondary circuit only. A separate switch must be provided for the primary.

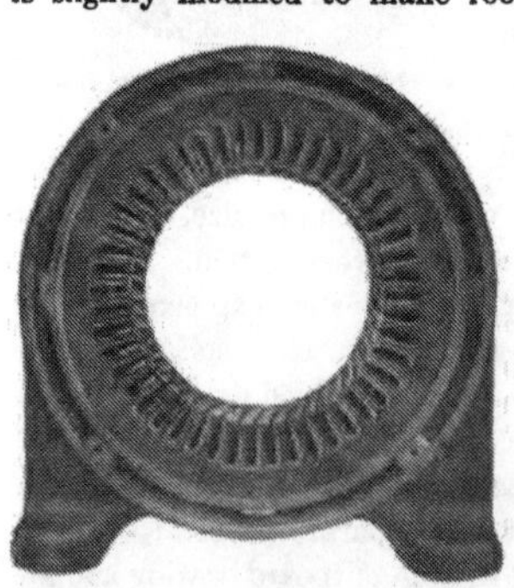

Fig. 175. Stator of Bell Automatic Single-Phase Motor

Bell Electric Motor Company. *Single-Phase Induction Motors.* The Bell Company makes a line of single-phase induction motors from ½ to 15 h. p. for 110 or 220 volts, 60 cycles. Their general appearance is shown by Fig. 174. The line current passes into the field or stator winding only, all currents in the armature being devel-

Fig. 176. Wound Armature of Bell Single-Phase Motor

oped by induction. The stator, Fig. 175, consists of punchings of the highest grade of laminated sheet iron, thoroughly annealed and slotted on the inner periphery to hold the field windings. These laminations are supported and protected by a light cast-iron frame carrying the feet of the motor. The bearings are of the best phosphor bronze and the shafts of high carbon steel. All shafts have oil slings so as to return to the reservoirs the oil distributed to them by revolving rings. By turning the end plates 90 or 180 degrees, these motors may be mounted on side wall or ceiling. The armature, Fig. 176, which is wound in a similar manner to those in direct-current motors, has a commutator and brushes which, being short-circuited

Fig. 177. Exploded View o fBell Short-Circuiting Device

on themselves, allow great starting torque with small starting current. While starting, therefore, the machine is a repulsion motor.

When the armature has attained full speed, all of the windings are short-circuited. The windings have now become the equivalent of a squirrel cage and the motor runs as a single-phase induction machine. The details of the short-circuiting device are shown in Fig. 177. After the motor has obtained nearly full speed, the commutator segments are entirely short-circuited by the copper ring, actuated by the centrifugal force of the weights. When the motor is stopped, the copper ring is pushed from the commutator segments by means of the steel spring and assumes its starting position. The direction of rotation may be reversed by simply loosening the set

Fig. 178. Stator of 5 H. P. Polyphase Induction Motor
Courtesy of Burke Electric Company

screw and rotating the rocker arm carrying the starting brushes to a new indicated position.

Burke Electric Company. *Polyphase Induction Motors.* The Burke Electric Company makes a line of induction motors of a very rugged and substantial frame, in many sizes, from ½ to 100 h. p., for the standard voltages and frequencies. The stator or primary member, Fig. 178, comprises the usual slotted core of laminated steel, mounted in a massive cast-iron housing and equipped with form-wound and individually insulated coils. The core laminations are keyed to the housing in order to avoid the use of bolts through the core, which entail eddy-current losses. The slots in the core are partly closed because this construction increases the effective area of the air gap, thus decreasing the gap reluctance and consequently the magnetizing current, and increasing the efficiency and the power factor. Radiation is secured on the outer surface of the core by a notched construction of the laminations. A thorough and uniform ventilation is produced by the revolving vanes on the rotor, which force currents of air against and up through the vent spaces, around the stator core, and through and around the windings. The rotor core disks are assembled with the conductor slots progressively twisted out of line just enough to make the completed slot diagonal, as illustrated in Fig. 179. This reduces the flux pulsations in the air gap, which tends to reduce iron losses and magnetic humming. The rotor conductors of the squirrel-cage type consist of one-piece closed loops punched out of copper sheet in the form shown at the top of

Fig. 179. Rotor of Burke Induction Motor Showing Diagonal Slots

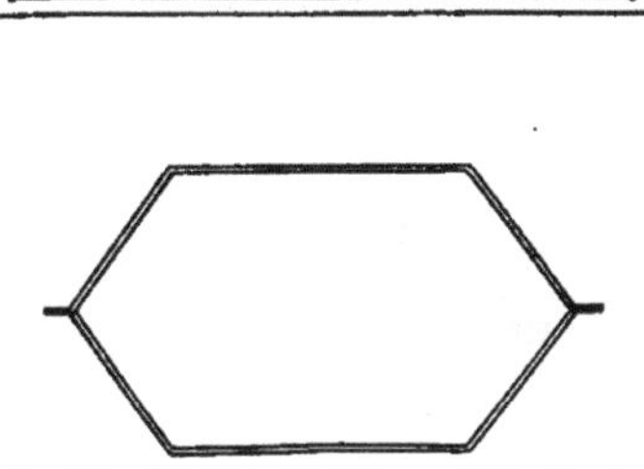

Fig. 180. Burke Rotor Bar Before and After Forming

Fig. 180, and afterward pulled out into shape, as shown at the bottom of the same illustration. The two straight sides or parallel sides of the loop are separated to a distance approximately equal to the pole pitch of the stator and it is, therefore, unnecessary to connect the ends of the various coils together; each one forms its own closed circuit. The slip-ring type of motor, Fig. 181, is, of course, equipped with a polar winding, the terminals of which are connected to collector rings. This form is used for variable-speed service and for conditions requiring high starting torque with moderate starting current. The bearings are mounted in the usual bonnets or brackets, fitted to the face and bore of the stator housing. The journal sleeves are cast-iron shells with babbitt linings, and the oil rings consist of a number of thin metal disks. Motors of 5 h. p. and under are not supplied with starters, but those of larger power employ starters of the resistor or auto-transformer types.

Fig. 181. Burke Rotor for Variable Speed-Induction Motors Showing Construction of Rotor and Slip Rings

Century Electric Company. *Single-Phase Induction Motors.* Fig. 182 shows the parts of a single-phase motor made by the Century Company. This line is built for all frequencies between 25 and 140 cycles and all voltages between 100 and 250 volts. As small power motors they are from $\frac{1}{50}$ to $\frac{1}{6}$ h. p.

Fig. 182. Exploded View of Century Split-Phase Induction Motor

Certain sizes are also used as a complete line of fan motors. They are built in two types: clutchless and clutch. The clutchless type is designed to develop a starting torque equal to full-load running torque but requires four or five times the full-load current to do it. It is well adapted for driving fans, blowers, and centrifugal pumps.

The clutch type will stand a heavier load but requires the same starting current. When the circuit is closed, the rotor starts to revolve on the shaft; when it reaches a certain speed, a three-piece centrifugal clutch expands and engages the clutch disk fastened to the shaft. The rotor is of the well-known squirrel-cage type; bare copper bars are imbedded in the laminations, securely fastened mechanically, and then soldered to bare copper rings on each end of the rotor. The field is built up of thin laminations wound with form-wound coils that are thoroughly impregnated with oil and

Fig. 183. Parts of Single-Phase Repulsion Induction Motors
Courtesy of Century Electric Company

moisture-resisting paint and then pressed into the motor frame. The split-phase or starting coils are in circuit only during the period of acceleration. When the motor reaches nearly full speed, a centrifugal switch opens this circuit.

Repulsion Induction Motors. A second line of motors in sizes from $\frac{1}{8}$ to 40 h. p. (Fig. 183), are single-phase, constant-speed, repulsion induction motors; that is, while running they are single-phase induction motors but start as repulsion motors. On reaching full speed, the governor weights are expanded and move a device which short-circuits every commutator bar to one common ring of high conductivity and at the same time releases the tension on the carbon brushes and pushes them back away from the commutator. When the motor is stopped or slows down to a low speed, the governor device automatically returns to its starting position.

Crocker-Wheeler Company. *Polyphase Induction Motors.* The Crocker-Wheeler Company manufacturers full lines of polyphase

Fig. 184. Assembled Stator Core and Housings of Form Q Induction Motor
Courtesy of Crocker-Wheeler Company

induction motors for 60 and 25 cycles, at the standard voltages. They embody the usual features of cast-iron frame, cast-steel laminations for stator core, form-wound interchangeable coils in stator winding, laminated core on rotor, with squirrel-cage or polar winding as desired, and usual methods for obtaining cool running. Figs. 184 and 185 show a stator and a rotor of these machines.

Some of their special features not employed in other makes are that the stator slots are first made open, allowing plenty of space for inserting form-wound, well-insulated coils, and are then closed by magnetic wedges which give mechanical strength and protection to the coils, at the same time giving all the electrical advantages of a closed-slot construction. The action of these magnetic wedges can be seen from the following considerations, as shown in Figs. 186 and 187. Fig. 186 shows an outline of the teeth with the air gap between them and the rotor exaggerated to show the path of the magnetic flux. It will be noted that the teeth *A* are fitted with magnetic

Fig. 185. Rotor of Crocker-Wheeler Induction Motor

bridges *E* and the teeth *B* are fitted with the wood-stick wedges *P*, according to the usual open-slot practice. Considering the teeth *A*, it will be noted that the magnetic bridge forms a shoe which increases the area from which the magnetic flux is distributed. It will be seen that this enables the flux to travel straight across the air gap, as shown. By observing the teeth *B*, it will be seen that the lines of force radiating from the comparatively smaller end of the teeth do not confine themselves to a straight path, but spread out fan-wise, many of them taking a diagonal path across the air gap. This makes the average path longer than in the case of the teeth *A*. The magnetic bridge, therefore, has the same effect as shortening the air gap and improves the power factor. In order to prevent flux leakage across the ends of adjacent teeth, the magnetic bridges are divided in the middle by a long slot *F*, as shown in Figs. 187 and 188. This slot is many times wider than the air gap between the teeth and the rotor and practically prevents leakage of flux in this

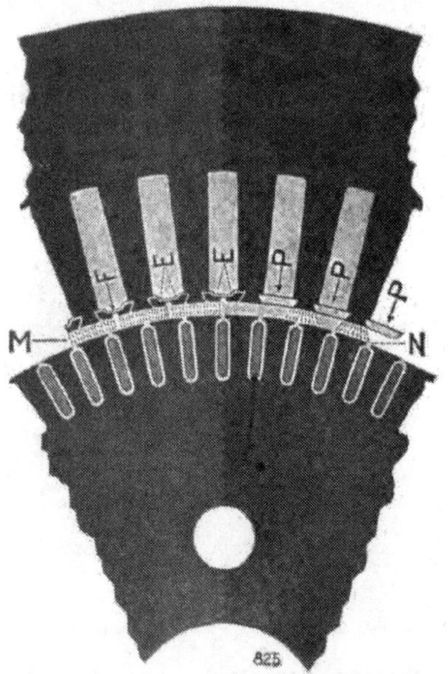

Fig. 186. Details of Magnetic Wedges for Crocker-Wheeler Induction Motors

direction. The connections by which the two halves of the bridge are held together are of such small cross section and are so highly saturated with magnetic flux that they need not be considered in connection with flux leakage. In order that this magnetic bridge *E* may not act as a short-circuit path between core laminations, it is insulated from the tooth *A* by the thin sheet of insulation *K*, Fig. 187. This insulation is effective for this purpose but offers very little resistance to the magnetic lines traveling across the air gap. Where wood wedges are used, pulsations of high frequency occur in the flux in the teeth, causing eddy-current losses. By the use of the magnetic bridges a better path for the flux is provided, thus largely avoiding the eddy-current effect and increasing the efficiency of the motor. Another feature is an improved type of end rings which connect the rotor bars in the squirrel-cage winding. Where continuous rings or butt-joint rings are used for this purpose, it is found that the current concentrates in the ring nearest the core. This increases the resistance and, by causing local heating, may melt the soldered connections. By the spiral arrangement of the bars shown in Fig. 189, these troubles are avoided and the electrical resistance of all bars made equal. The bars in the rotor are proportioned to give a moderately high resistance, resulting in a good starting torque. This result is accomplished without loss of efficiency, owing to the gain realized by the use of the magnetic bridges. In their slip-ring motors, the brushes are made of a composition of carbon which is submitted to an electroplating process while in a pulverized state, and each of the small particles is given a copper coating. After this process, the material is compressed into a solid brush.

Fig. 187. Section of Magnetic Bridge in Place
Courtesy Crocker-Wheeler Company

Fig. 188. Magnetic Bridge

The smaller sizes below 5 h. p. may be started by simply closing a switch connecting the stator windings to the line. For sizes of 30 h. p. or under, the Crocker-Wheeler Company supplies a switch by means of which the machine is started with its stator coils connected in star, or Y-fashion, and the motor takes about one-third of the current that it would, if connected directly across the line. As the speed increases, a movement of the handle makes the change to delta (Δ) connection, the normal running condition. With this star-

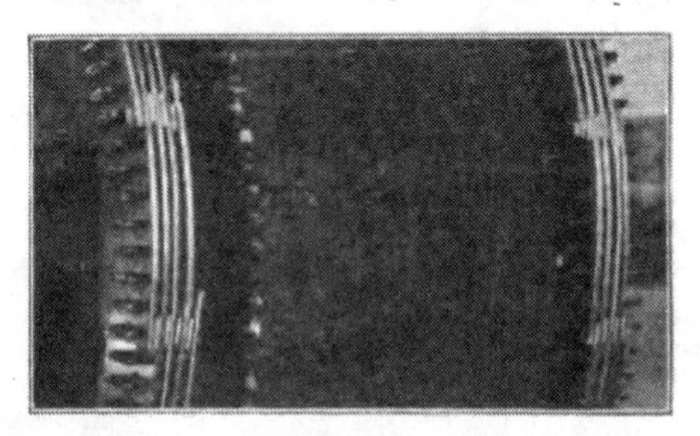

Fig. 189. Crocker-Wheeler End-Ring Connections

delta method, the effect at starting is the same as if about 58 per cent of normal voltage were applied to the motor terminals. Three-phase motors above 35 h. p. capacity and, of course, all two-phase motors are not adaptable to the star-delta method of starting, but are started by the use of an auto-transformer connected through a starting switch to the stator windings of the machine. By means of the transformer a low voltage is first applied to the terminals of the motor to start it, and after the machine has increased its speed, the voltage is raised in steps to normal value by means of a controller.

Fig. 190. Line of Single-Phase Induction Motors
Courtesy of Emerson Electric Manufacturing Company

The Emerson Electric Manufacturing Company. *Single-Phase Induction Motors.* The Emerson Company makes a line of single-phase induction motors designed for all frequencies from 25 to 133 in sizes from $\frac{1}{20}$ to $\frac{1}{2}$ h. p. and operating on 100 to 115 volts.

These same designs do for fan motors, small power motors, and for driving exhaust fans. These motors, the general appearance of which is shown by Fig. 190, are made in two styles, clutch and clutchless. The clutch type is provided with an internal automatic clutch that allows the rotor or armature to attain stable speed before engaging the shaft or the load of the driven machinery. In addition

Fig. 191. 50 H. P. Form K Induction Motor
Courtesy of General Electric Company

to the main field or stator winding of the motor, and connected in multiple with the main winding, is an auxiliary or starting winding placed midway between each main coil. The machines, therefore, are split-phase while starting. These starting coils are only connected in circuit during the period of acceleration, being automatically open-circuited by a centrifugal switch when the rotor attains its proper speed. The frames have housings cast with ventilating openings, allowing free ventilation through the field of the motor. The motor field or stator consists of laminations of specially selected steel. The punchings are slotted and the stator windings are carefully insulated and imbedded in the slots. The rotor or armature is also built up of laminations with heavy copper conductors and is of the ordinary squirrel-cage type.

General Electric Company. *Polyphase Induction Motors.* The G. E. Company builds complete lines of polyphase induction motors from ½ to 6000 h. p. for standard frequencies of 25, 40, and 60 cycles at the standard voltages of 110, 220, 440, 550, and 2200 volts for two-phase or three-phase operation. Their intermediate sizes, built in three types, employ the method of construction called the skeleton frame, as shown in Fig. 191. The three different types employ the same stator but differ in the rotors employed.

The stator is built up of circular laminations, keyed to the frame ribs and held together at each end by iron rings securely fastened to the frame. Besides having the usual ventilating ducts,

Fig. 192. Soldered Form of End-Ring Construction for Form K Rotor
Courtesy of General Electric Company

the outer circumference of the laminations is almost entirely exposed to the air, thus increasing the cooling surface. The interchangeable stator windings are form-wound and placed in open slots which are used in all but the smaller sizes.

The rotor is built up of annular punchings, dovetailed to the spider arms. Bolts passing through solid end rings underneath the punchings clamp the laminations together. The partly closed slots for the conductor bars assist in holding the windings in place.

Fig. 193. View of General Electric Form K Rotor Showing Welded Type of End-Ring Construction

The different rotor windings give rise to the three different types of these machines, forms K, L, and M. Form K employs a squirrel-cage winding, consisting of bars laid in the core slots and short-circuited at the ends by copper rings. For the smaller sizes these rings are thin but of considerable radial depth and are held apart by spacing washers. They have rectangular holes punched near their outer peripheries through which the rotor bars pass. Lips are formed in the rings of ample area, to which the bars are thoroughly soldered, as shown in Fig. 192. In the larger frames, on account of the difficulty of providing multiple soldered rings of sufficient cross section, a welded ring construction, as shown in Fig. 193, is employed. This consists of a cylindrical copper ring of ample width and cross section placed beneath the bars at each end of the rotor. Short radial bars are welded to the edges of these rings and to the rotor bars, thereby making a good electrical contact and rendering the structure mechanically secure. These rings improve the ventilation of the motor, when running, by drawing in a current of air and forcing it through the ends of the stator coils and ventilating ducts.

To reduce the current at starting and increase the torque, the form L motor, unlike the form K, is provided with a wound rotor, that has a starting resistance and switch located on the shaft within the rotor. Form L motors are used in preference to those of the form K type where the voltage regulation of the system is of importance. The starting resistance in motors up to about 35 h. p. consists of cast-iron grids enclosed in a triangular frame that is bolted to the end plates holding the rotor laminations together, and is short-circuited by sliding laminated spring metal brushes along the inside surface of the grids. The brushes are supported by a metal sleeve sliding on the shaft and operated by a lever secured to the bearing brackets. A rod passing through the end of the shaft operates the short-circuiting arrangement. Intermediate size frames use as resistances brass grids arranged in three sets 120 degrees

Fig. 194. Riveted-Frame Induction Motor with Sliding Base
Courtesy General Electric Company

apart. These are bolted to end plates holding the rotor laminations together and are short-circuited by sliding laminated copper brushes along the inside surface of these grids. These brushes are supported by a yoke sliding on the shaft and controlled by a lever. For the largest frames, cylindrical coil resistances of German silver wire wound on edge are used. These coils are bolted 120 degrees apart to bosses on the spider hub and are clamped together by a ring on their front end. Two laminated metal brushes bear directly on each of these resistances and are supported on a yoke sliding on the shaft.

The form M rotor is similar in construction to form L except that collector rings and controller are necessary because of the change in the location of the resistance which, with the controller, is placed external to the motor. Two or more carbon brushes are used for each collector ring. The smaller sizes from ½ to 15 h. p. are supplied by a line employing what is known as riveted-frame construction, illustrated in Fig. 194. The core frame commonly employed to hold the stator punchings has been entirely eliminated by securely clamping the laminations between two cast-iron flanges or end plates while under hydraulic pressure. The punchings are of sufficient depth to insure rigidity in the stator and, owing to their exposed surface, the radiation is improved. Form K and form L motors are constant-speed. With the use of suitable external resistances and compensators the form M type are either variable-speed or constant-speed motors where the starting conditions are particularly severe. Fig. 195 shows a three-phase variable-speed, form M, riveted-frame induction motor.

Fig. 195. General Electric Variable-Speed Riveted-Frame Induction Motor

Fig. 196. Type KS Single-Phase Induction Motor
Courtesy of General Electric Company

A line of multi-speed induction motors are built for three-phase circuits only. They are built with the standard riveted-frame construction and can be designed for operation requiring constant torque or for constant horsepower. They run at 600, 900, 1200, and 1800 r. p. m., the four speeds being obtained by changing the polar groupings of the stator coils.

Standard Single-Phase Induction Motors. The single-phase induction motors of General Electric manufacture are built in two lines, type KS and type RI. The standard type KS machines illustrated in Fig. 196, are built in sizes from 1 to 15 h. p. for 60 cycles and run at 1800 and 1200 r. p. m. The stators have symmetrical three-phase windings with form-wound coils placed progressively in the slotted punchings. The rotor is of smooth core, squirrel-cage,

Fig. 197. Type RI Single-Phase Repulsion Induction Motor
Courtesy of General Electric Company

high-resistance type consisting of soft steel disks assembled upon a steel sleeve, the disks being slotted near the circumference to retain the bar winding which extends beyond the core at both ends where it is permanently connected to heavy short-circuiting rings. The rotating member is supported upon the shaft by a special lining bearing interposed between the steel assembly sleeve of the rotor and the shaft. Rotor acceleration is accomplished by means of a starting box containing both resistance and reactance units and operated in much the same manner as the well-known direct-current motor rheostat. The rotor revolves freely on the shaft until about 75 per cent of rated speed is reached, when the load is picked up by the automatic action of a centrifugal clutch that rigidly engages an outer shell keyed directly to the shaft.

Compensated Repulsion Induction Motors. Type RI machines, illustrated in Fig. 197, are really compensated repulsion induction motors having a combination of series and shunt characteristics and capable of operation above or below synchronous speeds. RI motors are built in sizes up to 15 h. p. in the riveted-frame form. The field consists of slotted laminations assembled between end flanges and wound with two windings; a main winding of the distributed concentric type, and a compensating winding which is either the center portion of the main winding or a separate winding concentric therewith, depending upon the size of the frame used. The armatures are built up of selected sheet-steel laminations in which the coil slots are punched before being assembled. In sizes up to and including 5 h. p. the laminations are built up directly on the shaft, the larger sizes employing cast-iron spiders held in place by retaining rings and cast-iron core heads. The armature winding is of the series type, the smaller sizes being form-wound, while the larger employ bar windings. The commutators are made from the best grade, hard-drawn, high conductivity copper segments, insulated with selected mica, slotted to reduce the brush contact resistance and friction. To secure proper ventilation, in addition to the usual ducts between laminations of the stator, the armature shaft is fitted with a rigid fan. The brushes are in two different sets; one, the energy brushes; the second, the compensating, connected to the compensating field winding.

The action of the machine is as follows: In the straight repulsion motor, to secure the necessary starting torque, a direct-current armature is placed in a magnetic field excited by an alternating current and short-circuited through brushes set with a predetermined angular relation to the stator. To further improve the operating characteristics of the plain repulsion motor, a second set of brushes, the compensating set, is placed at 90 electrical degrees from the main short-circuiting brushes, the energy brushes, and is connected to the compensating field. This field is auxiliary to the main field and impresses upon the armature an electromotive force which is in angular and time phase with that generated by the main field. This improves the phase relation between current and voltage (resulting in high power factor), serves to restrict the maximum no-load speed, and also permits, when desired, a slight increase over synchronous values.

RI motors can also be built reversible and for varying and adjustable speed operation. The reversibility is accomplished by the addition of an auxiliary winding spaced 90 degrees from the main field winding and connected in series with it. By reversing the relative polarity of the two windings by means of a reversing switch, the direction of rotation is changed in a simpler manner than if the reversal were secured by mechanically shifting the radial position of the brush holder yoke. The varying-speed brush-shift motor is obtained by using a slight modification in the windings and brush rigging. To a grooved ring on the movable brush yoke is attached a flexible steel cable supported and guided in any desired direction by a small grooved pulley. The terminals of this cable are fastened respectively to the controller handle and the movable brushes.

Fig. 198. General Electric Single-Phase Motor with Wound Field and Drawn Shell

RI adjustable-speed motors allow a speed range of 2 to 1, about half this range being above synchronous speed and half below. This is obtained by modifying the windings and employing transformers whose primaries are excited by the line circuit. The secondaries of these transformers are divided into two sections; the first, or regulating circuit, is placed across the energy brushes; the other section is connected in series with the compensating winding.

Small Size Induction Motors. A line of very small motors from $\frac{1}{50}$ to $\frac{1}{4}$ h. p. employs the drawn shell construction of the similar size direct-current motors. They are single-phase induction motors with squirrel-cage armatures and employ the split-phase method of starting. A small brush carried by an insulated brush holder mounted on the motor cap acts as a switch to open the starting phase of the field windings when the armature reaches a predetermined speed. Fig. 198 shows a wound field of one of this type.

Synchronous Motors. The General Electric synchronous motors

Fig. 199. 187 K. V. A. Synchronous Condenser with Exciter
Courtesy of General Electric Company

are exactly the same as their various lines of alternators. In the smaller sizes their self-excited revolving-armature alternators, shown

Fig. 200. Rotor of General Electric Synchronous Condenser

in Fig. 149, are employed. These are then provided with grids on the poles and the field winding may be broken up by means of switches to keep down the induced voltages during starting, when this is done from the alternating-current side. This line of generators, when so modified, are also used as synchronous phase-modifiers, or synchronous condensers. They are so-called because a synchronous motor, when used as a synchronous condenser, has the property of altering the phase relation between voltage and current, the direction and extent of the displacement being dependent on the field excitation of the synchronous condenser. It can be run at unity-power factor and minimum current input, or it can be over-excited, and, thereby, take leading current which compensates for the induc-

Fig. 201. Triumph Electric Company Phase-Wound Motor

tive load on other parts of the system. Fig. 199 shows a machine of this type and Fig. 200 shows the grid winding carried in pole pieces of the rotor. This acts as a squirrel-cage winding during starting and minimizes hunting during running.

Triumph Electric Company. *Polyphase Induction Motors.* The Triumph Company manufactures a complete line of polyphase induction motors in sizes up to and including 200 h. p., wound for 110, 220, 440, 550, and 2300 volts for two-phase and three-phase circuits and for 25, 40, and 60 cycles. They can be arranged for any desired mounting, including the vertical-shaft type. The stator core is built from punchings of a special non-ageing steel, thoroughly japanned. The thoroughly insulated coils are form-wound and held firmly in position in the stator slots by wedges. Besides the insulation around the coils, the slots are also lined with insulating materials. The squirrel-cage rotor is built up of thin sheet-steel laminations, thoroughly japanned, and clamped together by heavy malleable-iron end-plates. Semi-enclosed slots are punched in the outer periphery to receive the windings and hold them in place against the action of centrifugal force. These conductors are set on edge and are riveted and soldered into heavy resistance rings of ample section. These rings are punched to receive the conductors in such a manner that there is an unbroken strip of metal completely surrounding each conductor. The short-circuiting rings are set some distance from the ends of the core so that the rotor bars between the core and the ring act as vanes and force large volumes of air through the coils and ventilating openings in the stator frame. For adjustable-speed work and for extremely heavy starting duty, phase- or wire-wound rotors are employed, as illustrated in Fig. 201. The rotor circuits are completed by connecting to the collector rings mounted upon the shaft suitable external resistances.

Wagner Electric Manufacturing Company. *Polyphase Induction Motors.* The Wagner Company builds complete lines of induction motors for standard frequencies at 110, 220, and 440 volts, wound either two-phase or three-phase. The stator forms the field and the usual construction is employed. The rotor is the armature and is built in the usual squirrel-cage or wound-rotor type, the squirrel-cage rotor being employed for constant speed. In order to prevent any possibility of the rotor bars shifting lengthwise and thereby unbalancing the rotor, al squirrel-cage end rings are shouldered on the armature flanges. The wound rotors with their slip rings and external resistances are used for variable-speed motors or constant-speed motors, wherever the load at starting is heavy, so as to avoid voltage disturbance resulting from heavy starting currents. In these lines the angular position of the end plates on the frames may be shifted so as to permit the installation of the motors in any position on floor, wall, or ceiling.

Another form of polyphase induction motor put upon the market by this company is their type BW. It is built in complete lines of three different speeds 1800, 1200, or 900 r. p. m. for 60 cycles, at

110, 220, or 440 volts in sizes from 3 to 50 h. p. and wound either two-phase or three-phase. These machines in outward appearance are like the single-phase machines built by this company. The stator or field is built according to the usual standard construction of polyphase squirrel-cage induction motors of other makes. The rotor or armature has a distributed winding, tapped to a vertical commutator in such a way that, by short-circuiting all the segments, the winding is converted into one of a squirrel-cage type. This is accomplished by a centrifugal device that acts shortly before the motor reaches full speed. These motors are constant-speed poly-

Fig. 202. Single-Phase, Unity Power Factor Motor
Courtesy of Wagner Electric Manufacturing Company

phase motors suitable for practically all installations in which ordinary squirrel-cage motors can be used and will also take the place of the usual wound rotor type for any purpose not requiring speed variation. They have all of the advantages of the wound rotor during starting.

Constant-Speed Single-Phase Motors. A novel form of constant-speed single-phase motor, shown in Fig. 202 and known as type BK, is built by the Wagner Company in sizes up to 15 h.p. Standard methods of mechanical construction are followed. The windings on both the stator and the rotor, as well as the principles of operation, are different from any other machine upon the market. In the stator or field construction two windings are used instead of one. The main stator winding, shown as winding *1*, Fig. 203, produces the initial field magnetization. The auxiliary winding *2* controls the power factor, or compensates the motor. The main structural novelty is in the rotor, the construction of which is clearly indicated in Fig. 204. Here again two windings are employed. The main or principal winding *4* is of the usual well-known squirrel-cage type and occupies the bottom of the rotor slots; the second or auxiliary winding *3* is of the usual commuted type, and is connected to a standard form of horizontal commutator and occupies the upper portion of the rotor slots. Between the two is placed a magnetic separator in the form of a rolled-steel bar. Two sets of brushes are provided, as indicated in Fig. 203. The main pair of brushes *5-6* is placed in the axis of the main stator winding *1* and is short-circuited. The auxiliary pair of brushes *7-8* is placed at right angles to the axis of the main stator winding and is connected in series with it. The auxiliary stator winding *2* is permanently connected to one auxiliary brush *7*, and is connected to the other auxiliary brush *8*, after starting by means of a small automatic centrifugal switch provided on the outer extension of the rotor shaft (See left end of motor shaft, Fig. 202) and which operates after a sufficient speed has been reached. This is represented by switch *9* in Fig. 203. This system of windings and connections is for the purpose of accentuating at starting the effect of the squirrel cage along the axis *5-6* of the main stator winding *1*, while suppressing it as far as possible along the axis *7-8* at right angles to *1*. The magnetic separator placed above the squirrel-cage winding *4* tends to suppress the effect of that winding along all axes,

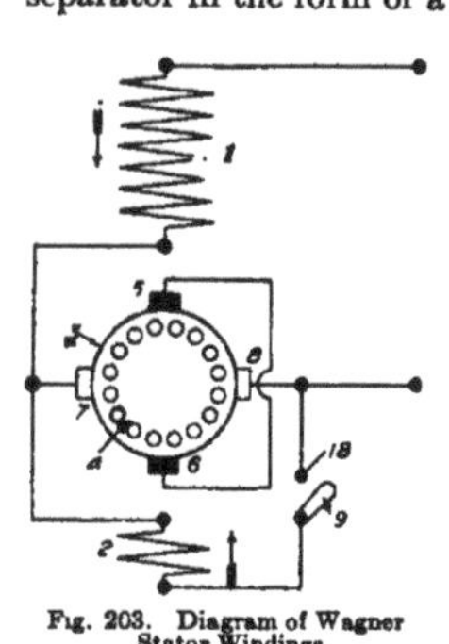

Fig. 203. Diagram of Wagner Stator Windings

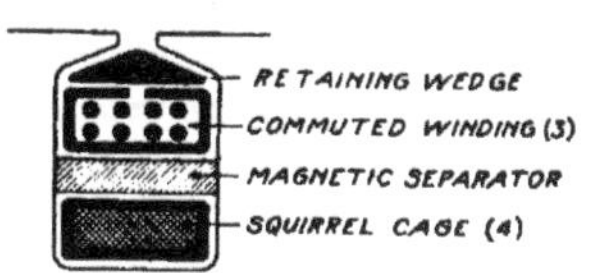

Fig. 204. Section Showing Construction of Wagner Rotor

by making said winding less responsive to outside inductive effects. The influence, however, of this separator is nullified along the axis of the main stator winding by the presence of the short-circuited brushes *5-6*, while no means are provided for nullifying its effect along the axis at right angles to that of the main stator winding. Thus the main stator winding *1* will be able to induce heavy currents in both rotor windings because of the short-circuited brushes in the axis *5-6* in spite of the magnetic separator; while the rotor winding *3*, connected in series with *1*, will not be able to produce heavy currents in the squirrel-cage winding *4* along the axis *7-8* because of the magnetic separator between *3* and *4* shunting the inducing magnetic flux.

At starting, switch *9* of Fig. 203 is open, the commuted winding *3* along the axis *7-8* being connected in series with the main stator winding *1* across the mains. The winding *1* induces a large current in the rotor windings *3* and *4* along the axis *5-6*, and the winding *3* produces a large flux along the axis *7-8*, the motor starting as a series machine. As the motor speeds up, the squirrel cage gradually assumes those functions that it performs in the ordinary single-phase motor and produces a magnetic field of its own along the axis *7-8*. Since the magnetizing currents circulating in the bars of the squirrel cage of a single-phase induction motor at synchronism are double the frequency of the stator currents, the fluxes they produce must be of double frequency. Now the magnetic separators are so proportioned of solid steel that while they form sufficiently effective shunts for the fluxes of line frequency induced from the stator, they are quite ineffective as shunts for the double-frequency fluxes produced by the rotor.

As far as the squirrel cage is concerned, the effect of the magnetic separator diminishes with increasing speed and at synchronism the machine operates practically in the same manner as if the separator did not exist at all. This form of motor under running conditions has a power factor leading at light loads and practically unity from half load to fifty per cent overload. The employment of the squirrel-cage winding in combination with the commuted winding secures a very small change in speed from no load to considerable overload, the speed being slightly above synchronous speed at light loads. The squirrel cage also prevents the motor from racing or running away.

Westinghouse Electric and Manufacturing Company. *Squirrel-Cage Induction Motors.* The new line of Westinghouse type CS squirrel-cage induction motors possesses several new features, among which are the extensive use of pressed steel in the construction and rotors with cast-on short-circuiting rings. These motors are put upon the market in all commercial sizes from 1 to 200 h. p. for the standard frequencies and voltages. A 10 h. p. machine is shown in Fig. 205. Pressed steel imparts great mechanical strength and is very uniform in structure, hence a motor of given weight can be made with more active material than motors of corresponding capacity in

Fig. 205. Type CS Squirrel-Cage Induction Motor
Courtesy of Westinghouse Electric & Manufacturing Company

cast-iron frames. In these motors, rolled steel is used in the frames of the sizes above 20 h. p., as well as in the end plates of the smaller sizes and in the feet and slide rails of all sizes. Above 5 h. p. the form-wound stator coils are laid in open slots. In all sizes the rotor bars are insulated with a special cement, which is moisture-proof and will withstand a high degree of heat and large mechanical stress. In motors above 15 h. p. the bars are connected electrically and mechanically by casting the short-circuiting rings around the ends. The bearings are protected from dust by a cap on the front end and by felt washers between metal rings on the pulley end.

Phase-Wound Slip-Ring Motors. Another line called type HF are phase-wound slip-ring motors, as shown in Fig. 206. They are made in capacities ranging from 5 to 200 h. p. for two-phase or three-phase circuits of 25 and 60 cycles; small motors are made for voltages up to 550 and large motors for voltages up to 2200. The

Fig. 206. Type HF Polyphase Induction Motor with Bed Plate
Courtesy of Westinghouse Electric & Manufacturing Company

frame is a one-piece cylindrical iron casting. The stator core is built up of sheet-steel laminations, enameled before assembling, clamped between cast-iron end plates, and keyed or dovetailed to lugs cast inside the frame as shown in Fig. 207. The stator windings are form-wound coils of insulated wire or strap. For the lower voltage machines a semi-enclosed insulated slot is used, while for frames for high voltages open slots are employed. The rotor laminations are enameled, assembled on a cast-iron spider, and clamped between

Fig. 207. Primary of Westinghouse Type HF Motor Without Winding

Fig. 208. Rotor of 50 H. P. Type HF Westinghouse Motor Showing Collector

iron end plates; the spider is then pressed on the shaft and keyed. Wherever necessary, the end plates are cast with extensions to support the ends of the rotor coils. The rotor slots are skewed, as shown in Fig. 208, in all except the largest motors. The rotor windings are three-phase, star-connected, and of insulated copper wire or strap placed in partly closed slots. The collector consists of three copper alloy rings assembled on a cast-iron bushing that is pressed on the motor shaft and keyed. One, two, or four carbon brushes per slip ring are used according to the size of the motor. When

Fig. 209. Westinghouse Type AR Single-Phase Repulsion Motor

called for, motors of 100 h. p. and larger are built with a device for directly short-circuiting the secondary windings shunting the current from the collector rings, brushes, and controller.

Small Single-Phase Repulsion Type Motors. The Westinghouse Company also builds smaller single-phase motors of the repulsion starting type. Their type AR motors illustrated in Fig. 209, are built in capacities of 2, 3, 5, 7½, and 10 h. p. for 60 cycles, 110 or 220 volts and synchronous speed of 1200 and 1800 r. p. m. The stator construction is shown by Fig. 210. The primary winding consists of laminations riveted together under pressure, pressed-steel end plates being riveted to the unit thus formed. This construction combines great strength, light weight, and ease of ventilation. The stator coils are thoroughly insulated. The secondary, or rotor, Fig. 211, has laminations with spacers for ventilating ducts riveted between end plates, and the unit thus formed is keyed to the shaft. The coils are made of strap copper and are pushed into the slots from the pulley end. The coils are held in place by fibre wedges and band wires. Each motor is equipped with a centrifugal switch, which short-circuits the rotor windings and releases the brushes. This is

Fig. 210. Stator of Type AR Repulsion Motor

located inside the rotor at the commutator end, and consists of a steel sleeve, a centrifugal governor, and a spring. The sleeve carries a short-circuiting coil which consists of a helical phosphor-bronze spring inside of which is a ring of flexible copper shunts. When the motor is at rest, the short-circuiting sleeve is pressed back into the rotor by the spring. When the motor speeds up, centrifugal force causes the governor weights to move outward, and the sleeve is forced forward. At nearly full speed, the short-circuiting coil is forced under the ends of the commutator bars and into very close contact with them, thus completely short-circuiting them. At the same time the end sleeve presses back the brush springs, and the brushes, being free to move away from the commutator, are pushed back by the end-play of the rotor.

Fig. 211. Rotor of Type AR Repulsion Motor

Split-Phase-Starting Induction Motors. A line of small power motors ranging from $\frac{1}{20}$ to $\frac{1}{4}$ h. p. called type DA are built for 110

Fig. 212. 600 K. V. A. Self-Starting Synchronous Condenser
Courtesy of Westinghouse Electric & Manufacturing Company

and 220 volts. They are single-phase induction motors, starting split-phase and having squirrel-cage armatures. They are furnished with and without automatic centrifugal clutches as desired. A line of alternating-current fan motors are also put upon the market by this company, being split-phase-starting induction motors for 40, 50, 60, and 133 cycles and series-wound motors for 25 and 60 cycles.

Synchronous Motors. For Westinghouse synchronous motors type E generators, shown in Fig. 160 are used. They are then, provided with the cage winding on the rotor (see Fig. 164) which makes the motor self-starting under light loads and tends to prevent hunting during operation. This line of generators when supplied with the cage winding are also employed as synchronous phase modifiers or synchronous condensers, as shown in Fig. 212.

ROTARY CONVERTERS

General Characteristics. The rotary converter used in conjunction with step-down transformers provides the most efficient, reliable, and simple means of converting alternating current into direct current. In its essential features it is similar to a direct-current generator with the addition of suitable collector rings connected to the armature winding at points having the proper phase relation. It combines in a single machine the functions of a synchronous motor and a direct-current generator. They are built single-phase, two-phase, three-phase, and six-phase and for 25 or 60 cycles. Those employed in connection with railway circuits have a standard direct-current voltage of 600 or 1200, while those used on three-wire lighting circuits have a direct-current voltage of from 240 to 300. Single-phase converters for use in battery charging, moving-picture work, telegraph and telephone systems, X-ray work, etc., usually have a direct-current voltage near 125 volts.

The ratio of the direct-current and alternating-current voltages in a rotary converter is practically a fixed quantity and this is a serious disadvantage where it is desirable to have the direct-current voltage at the machine increase with the load so as to keep it constant at some distant point. The methods employed for obtaining this change in voltage give rise to certain modifications in the design of the rotary converter and make necessary certain auxiliary devices.

Compound-Wound Rotary Converter. Fig. 213 shows a compound-wound rotary converter. Used in connection with external reactances, this type of machine is standard for railway work. The range in voltage obtainable is just about sufficient to take care of the drop in voltage from no load to full load between the generating station and the direct-current side of the rotary converter. This is due to the facts that adjusting the field strength of a rotary converter

Fig. 213. General Electric 600 Volt Compound-Wound Rotary Converter

changes the phase of the current just as in a synchronous motor, and also that a lagging alternating current passing through an inductive circuit causes a decrease in voltage while a leading current will cause a rise. Therefore, by placing sufficient reactance on the alternating-current side of a rotary converter the voltage at the collector rings can be varied by changing the field strength.

Regulating Pole Converter. Fig. 214 shows a General Electric regulating pole rotary converter. In this machine the variation of the voltage ratio is obtained, not by a variation of the impressed alternating voltage, but by varying the distributed flux under the poles, or, as it is usually called, by varying the field form of the converter. The field structure is divided into two parts, a main pole and a regulating pole; and the ratio between the direct and the alternating voltages can be readily varied by varying the excitation of the regulating poles, the only auxiliary apparatus required being a special field rheostat for controlling the exciting current. For

Fig. 214. General Electric Regulating Pole Rotary Converter

normal voltage the main pole only is excited, while in order to raise or lower the direct voltage, the regulating pole is excited so as to assist or oppose the effect of the main pole.

Shunt-Wound Converter with Synchronous Booster. Fig. 215 shows a General Electric shunt-wound converter with synchronous booster. This type of converter consists of a shunt-wound converter and an alternator with revolving field mounted on the same shaft as the converter armature. The armature of the alternator, or booster, is stationary and connected electrically in series between the supply circuit and the collector rings of the rotary converter.

Fig. 215. General Electric Shunt-Wound Rotary Converter with Synchronous Booster

Fig. 216. Westinghouse Rotary Converter Showing Direct-Current End with Oscillator and Speed-Limit Device

The booster field has the same number of poles as the converter. A change in the booster voltage will correspondingly change the

Fig. 217. Six-Phase Westinghouse Rotary Converter with Synchronous Regulator

alternating voltage impressed on the converter and this can be made so as to either increase or decrease the impressed voltage by strengthening or weakening the booster field.

Fig. 218. Converter Rotor with Synchronous Regulator, Armature, and Starting Motor Rotor
Courtesy of Westinghouse Electric & Manufacturing Company

Converter with Oscillator and Speed-Limit Device. Fig. 216 shows a Westinghouse rotary converter with oscillator and speed-limit device. Since a rotary converter will normally take a definite position in its bearings relative to the field, and revolve in a uniform plane without endwise oscillation, some device is necessary to produce a periodic axial movement of the armature shaft, assuring uniform wear of the commutator and the collector rings.

Converter with Synchronous Regulator. For the safe operation of rotary converters it is necessary that they be equipped with a device for automatically opening the circuit in case the speed becomes too high. Fig. 217 shows a Westinghouse rotary converter with a synchronous regulator. This company builds the synchronous regulator so that the armature is the revolving part, the armature windings of the regulator generator being connected in series

Fig. 219. Wagner Single-Phase Motor Converter

between the armature and the collector rings of the rotary converter. This method of construction is shown clearly in Fig. 218. Fig. 219 shows a single-phase rotary converter brought out by the Wagner Electric Manufacturing Company to be used for battery charging in automobile, telegraph, and telephone work.

MOTOR GENERATORS

Comparison with Rotary Converters. Motor generators, consisting of alternating-current motors and direct-current generators,

Fig. 220. 1200 KW Synchronous Motor-Generator Set
Courtesy of Crocker-Wheeler Company

are used for battery charging, for exciter sets in large alternating-current power stations, for railway sets, and for arc lighting sets. A motor-generator set may have decided advantages over a rotary

Fig. 221. 2300 Volt GE Synchronous Motor Connected to 550 Volt Generator

converter in special cases, as there is no electrical connection between the two sides of the system, an independent voltage adjustment is

Fig. 222. 13,200 Volt Synchronous Motor Connected to 1575 Volt DC Generator
Courtesy of General Electric Company

possible over a wide range, and the regulation is not so greatly affected by fluctuations in the supply circuit. Illustrations of this kind of motor generator are given in Figs. 220, 221, and 222. Fig.

Fig. 223. Low Voltage, 300 KW Direct-Current Motor-Generator Set
Courtesy of Westinghouse Electric & Manufacturing Company

223 shows a low-voltage 300 k. w. direct-current motor-generator set, manufactured by the Westinghouse Company.

Fig. 224. General Electric Frequency-Changer Set

Fig. 225. Westinghouse 1000 KW Frequency-Changer Set

Used as Frequency Changers. When the motor-generator set is composed of two synchronous machines designed to operate at different frequencies it is used as a frequency changer. For large operations in power service and railway service a low frequency has been generally adopted. There are, however, other classes of service, as lighting, for instance, where a higher frequency is desirable. A transformation from one frequency to another is easily effected by means of a motor-generator set made up of two alternating-current machines. The generator may be wound for any practical voltage, phase, and frequency, and the motor designed to operate from any commercial circuit. Figs. 224 and 225 illustrate examples of frequency changers.

VIEW OF GENERATOR FLOOR OF HALF THE MISSISSIPPI RIVER POWER HOUSE AT KEOKUK, IOWA

POWER STATIONS

INTRODUCTION

With the rapid increase of the use of electricity for power, lighting, traction, and electro-chemical processes, the power houses equipped for the generation of the electrical supply have increased in size from plants containing a few low-capacity dynamos, belted to their prime movers and lighting a limited district, to the modern central station, furnishing power to immense systems and over extended areas. Examples of the latter type of station are found at Niagara Falls, and such stations as the Metropolitan and Manhattan stations in New York City, and the plants of the Boston Edison Illuminating Company, etc.

The subject of the design, operation, and maintenance of central stations forms an extended and attractive branch of electrical engineering. The design of a successful station requires scientific training, extensive experience, and technical ability. Knowledge of electrical subjects alone will not suffice, as civil and mechanical engineering ability is called into play as well, while ultimate success depends largely on financial conditions. Thus, with unlimited capital, a station of high economy of operation may be designed and constructed, but the business may be such that the fixed charges for money invested will more than equal the difference between the receipts of the company and the cost of the generation of power alone. In such cases it is better to build a cheaper station and one not possessing such extremely high economy, but on which the fixed charges are so greatly reduced that it may be operated at a profit to the owners.

The designing engineer should be thoroughly familiar with the nature and extent of the demand for power and with the probable increase in this demand. Few systems can be completed for their ultimate capacity at first and, at the same time, be operated economically. Only such generating units, with suitable reserve capacity, as are necessary to supply the demand should be installed at first, but all apparatus should be arranged in such a manner that future extensions can readily be made.

Power stations, as here treated, will be considered under the following general topics:

Location of Station
Steam Plant
Hydraulic Plant
Gas Plant
Electric Plant
Buildings
Station Records

LOCATION OF STATION

The choice of a site for the generating station is very closely connected with the selection of the system to be used, which system, in turn, depends largely on the nature of the demand, so that it is a little difficult to treat these topics separately. Several possible sites are often available, and we may either consider the requirements of an ideal location, selecting the available one which is nearest to this in its characteristics, or we may select the best system for a given area and assume that the station may be located where it would be best adapted to this system. Wherever the site may be, it is possible to select an efficient system, though not always an ideal one.

The points that should be considered in the location of a station, no matter what the system used, are accessibility, water supply, stability of foundations, surroundings, facility for extension, and cost of real estate.

Accessibility. The station should be readily accessible on account of the delivery of fuel, of stores, and of machinery. It should be so located that ashes and cinders may easily be removed. If possible, the station should be located so as to be reached by both rail and water, though the former is generally more desirable. If the coal can be delivered to the bunkers directly from the cars, the very important item of the cost of handling fuel may be greatly reduced. Again, the station should be in such a location that it may readily be reached by the workmen.

Water Supply. Cheap and abundant water supply for both boilers and condensers is of utmost importance in locating a steam station. The quality of the water supply for the boiler is of more importance than the quantity. It should be as free as possible from impurities which are liable to corrode the boilers, and for this reason water from the town mains is often used, even when other water is available, as it is possible to economize in the use of water by the selection of proper condensers. The supply for condensing purposes should be abundant, otherwise it is necessary to install extensive cooling apparatus, which is costly and occupies much space.

Stability of Foundations. The machinery, as well as the buildings, must have stable foundations, and it is well to investigate the availability of such foundations when selecting the site.

Surroundings. In the operation of a power plant using coal or other fuels, certain nuisances arise, such as smoke, noise, vibration, etc. For this reason it is preferable to locate where there is little liability to complaint on account of these causes, as some of these nuisances are costly and difficult, or even impossible, to prevent.

Facility for Extension. A station should be located where there are ample facilities for extension and, while it may not always be advisable to purchase land sufficient for these extensions at first, if there is the slightest doubt in regard to being able to purchase it later, it should be bought at once, as the station should be as free as possible from risk of interruption of its plans. Often real estate is too high for purchasing a site in the best location, and then the next best point must be selected. A consideration of all the factors involved is necessary in determining whether or not this cost is too high. In densely populated districts it is necessary to economize greatly with the space available, but it is generally desirable that all the machinery be placed on the ground floor and that adequate provision be made for the storage of fuel, etc.

Cost of Real Estate. The location of substations is usually fixed by other conditions than those which determine the site of the main power house. Since, in the simple rotary-converter substation, neither fuel nor water is necessary, and there is little noise or vibration, it may be located wherever the cost of real estate will permit, provided suitable foundations may be constructed. The distance between substations depends entirely on the selection of the system and the nature of the service.

GENERAL FEATURES

Miscellaneous Considerations. Where low voltages are used, it is essential that the station be located as near the center of the system as possible. This center is located as follows:

Having determined the probable loads and their points of application for the proposed system, these loads are indicated on a drawing with the location of the same shown to scale. The center of gravity of this system, considering each load as a weight, is then found and its location is the ideal location, as regards amount of copper necessary for the distributing system.

Consider Fig. 1, which shows the location of five different

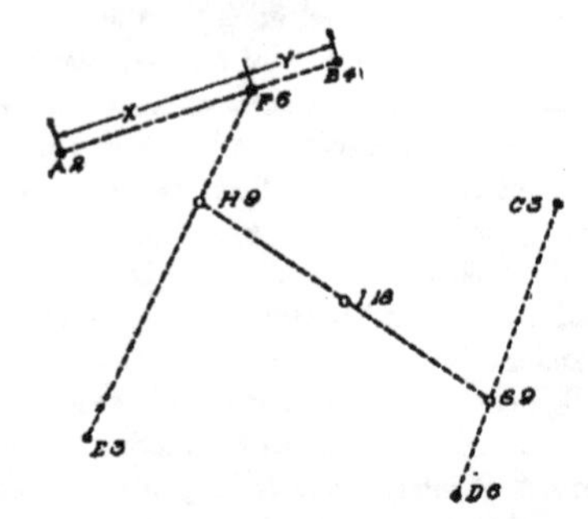

Fig. 1. Graphical Method of Locating Center of the System

loads, indicated in this case by the number of amperes. Combining loads A and B, we have

$$Ax = By \qquad x + y = a$$

Solving these equations, we find that A and B may be considered as a load of $A + B$ amperes at F. Similarly, C and D, E and F, and G and H may be combined giving us I, the center of the system. The amount of copper necessary for a given regulation runs up very rapidly as the distance of the station from this point increases. Where there are obstructions which will not permit the feeders to be run in an approximately straight line, the distance $A\ B$, etc., should be measured along the line the conductors must take.

Selection of System. General rules only can be stated for the selection of a system to be used in any given territory for a certain class of service.

For an area not over two miles square and a site reasonably near the center, direct-current, low-pressure, three-wire systems may be used for lighting and ordinary power purposes. Either 220 volts or 440 volts may be used as a maximum voltage, and motors should, preferably, be connected across the outside wires of the circuits. Five-wire systems with 440 volts maximum potential have been used, but they require very careful balancing of the load if the service is to be satisfactory. 220-volt lamps are giving good satisfaction; moderate-size, direct-current motors may readily be built for this pressure and constant-potential arc lamps may be operated on this voltage, though not so economically as on 110 volts, if single lamps are used. The new types of incandescent lamps in low candle-power units are not suitable for 220 volts. For direct-current railway work, the limit of the distance to which power may be economically delivered with an initial pressure of 600 volts is from five to seven miles, depending on the traffic.

If the area to be served is materially larger than the above, or distances for direct-current railways greater, either of the two following schemes may be adopted: (1) Several stations may be located in the territory and operated separately or in multiple on the various loads; or (2) one large power house may be erected and the energy transmitted from this station at a high voltage to various transformers or transformer substations which, in turn, transform the voltage to one suitable for the receivers. Local conditions usually determine which of these two shall be used. The alternating-current system with a moderate potential—about 2,300 volts for the primary lines—is now often installed for very small lighting systems.

The use of several low-tension stations operating in multiple is recommended only under certain conditions, namely, that the demand is very heavy and fairly uniformly distributed throughout the area, and suitable sites for the power house can readily be obtained. Such conditions rarely exist and it is a question whether or not the single station would not be just as suitable for such cases as where the load is not so congested.

One reason why a large central station is preferred to several smaller stations is that large stations can be operated more economically, owing to the fact that large units may be used and they can be run more nearly at full load. There is a gain in the cost of attendance, and labor-saving devices can be more profitably installed. The location of the power plant is not determined to such a large extent by the position of the load, but other conditions, such as water supply, cheap real estate, etc., will be the governing factors. In several cities, notably New York, Chicago, and Boston, large central stations are being installed to take the place of several separate stations, the old stations being changed from generating power houses to rotary-converter substations. Both direct-current low-tension machines—for supplying the neighboring districts—and high-tension alternating-current machines—for supplying the out, lying or residence districts—are often installed in the one station.

As examples of the central station located at some distance from the center of the load, we have nearly all of the large hydraulic power developments. Here it is the cheapness of the water power which determines the power-house location. The greatest distance over which power is transmitted electrically at present is in the neighborhood of 200 miles.

If a high-tension alternating-current system is to be installed, there remains the choice of a polyphase or single-phase machine as well as the selection of voltage for transmission purposes. As pointed out in "Power Transmission," polyphase generators are cheaper than single-phase generators and, if necessary, they can be loaded to about 80 per cent of their normal capacity, single-phase, while motors can more readily be operated from polyphase circuits. If synchronous motors or rotary converters are to be installed, a polyphase system is necessary. The voltage will be determined by the distance of transmission, care being taken to select a value considered as standard, if possible. Generators are wound giving a voltage at the terminals as high as 15,000 volts, but in many districts it is desirable to use step-up transformers for voltages above 6,600 on account of liability to troubles from lighting.

With the development of the single-phase railway motor, central stations generating single-phase current only, are occasionally built in larger sizes than previously, as their use heretofore has been limited to lighting stations.

Factors in Design. A few general notes in regard to the design of plants will be given here, the several points being taken up more in detail later.

Direct driving of apparatus is always superior to methods of gearing or belting as it is efficient, safe, and reliable, but it is not as flexible as shafting and belts, and on this account its adoption is not universal.

Speeds to be used will depend on the type and size of the generating unit. Small machines are always cheaper when run at high speeds, but the saving is less on large generators. For large engines slow speed is always preferable.

It is desirable that there be a demand for both power and lighting, and a station should be constructed which will serve both purposes. The use of power will create a day load for a lighting station, which does much to increase its ultimate efficiency and, as a rule, its earning capacity.

In addition to generator capacity necessary to supply the load, a certain amount of reserve, either in the way of additional units or overload capacity, must be installed. The probable load for, say three years, can be closely estimated, and this, together with the proper reserve, will determine the size of the station. The plant as a whole, including all future extensions, should be planned at the start as extensions will then be greatly facilitated. Usually it will not be desirable to begin extensions for at least three years after the first part of the plant has been erected.

Enough units must be installed so that one or more may be laid off for repairs, and there are several arguments in favor of making this reserve in the way of overload capacity, for the generators at least. Some of these arguments are:

Reserve is often required at short notice, notably in railway plants.

With overload capacity the rapid increase of load, such as occurs in lighting stations when darkness comes on suddenly, may more readily be taken care of.

There is always a factor of safety in machines not running to their fullest capacity.

Reserve capacity is cheaper in this form than if installed as separate machines.

As a disadvantage, we have a lower efficiency, due to machines not usually running at full load, but in the case of generators this is very slight.

TABLE I

Permissible Overload 33 Per Cent

	Machines added one at a time		Machines added two at a time		Machines added three at a time	
	No.	Size.	No.	Size.	No.	Size.
Initial installment	4	500	4	500	4	500
First extension	1	666	2	1000	3	2000
Second extension	1	888	2	2000	5	5000
Third extension	1	1183	2	4000	4	5000
Fourth extension	1	1577	4	4000		
Fifth extension	1	2103	8	4000		
Sixth extension	1	2804				

With an overload capacity of 33⅓ per cent, four machines should be the initial installment, since one can be laid off for repairs, if necessary, the total load being readily carried by three machines. In planning extensions, the fact that at least one machine may require to be laid off at any time should not be lost sight of, while the units should be made as large as is conducive to the best operation.

Table I is worked out showing the initial installment for a 2,000-kw. plant with future extensions. It is seen from this table that adding two machines at a time gives more uniformity in the size of units—a very desirable feature.

The boilers should be of large units for stations of large capacity, while for small stations they must be selected so that at least one may be laid off for repairs.

STEAM PLANT

BOILERS

The majority of power stations have as their prime movers either steam or water power, though there are many using gas. If steam is the power selected, the subject of boilers is one of vital importance to the successful operation of central stations. The object of the boiler with its furnace is to abstract as much heat as possible from the fuel and impart it to the water. The various kinds of boilers used for accomplishing this more or less successfully are described in books on boilers, and we will consider here the merits of a few of the types only as regards central-station operation.

The considerations are: (1) Steam must be available throughout the twenty-four hours, the amount required at different parts of the day varying considerably. Thus, in a lighting station, the demand from midnight to 6 A. M. is very light, but toward evening, when the load on the station increases very rapidly, there is an abrupt increase in the rate at which steam must be given off. The maximum demand can readily be anticipated under normal weather conditions, but occasionally this maximum will be equaled or even exceeded at unexpected moments. For this reason a certain number of boilers must be kept under steam constantly, more or less of them running with banked fires during light loads. If the boilers have a small amount of radiating surface, the loss during idle hours will be decreased.

(2) The boilers must be economical over a large range of rates of firing and must be capable of being forced without detriment. Boilers should be provided which work economically for the hours just preceding and following the maximum load while they may be forced, though running at lower efficiency, during the peak.

(3) Coming to the commercial side of the question, we have first cost, cost of maintenance, and space occupied. The first cost, as does the cost of maintenance, varies with the type and the pressure of the boiler. The space occupied enters as a factor only when the situation of the station is such that space is limited, or when the amount of steam piping becomes excessive. In some city-plants, space may be the determining feature in the selection of boilers.

Classification. Boilers for central stations may be classified as fire-tube and water-tube types. Of the former may be mentioned the Cornish, Lancashire, Galloway, multitubular, marine, and economic boilers. The Babcock and Wilcox, Stirling, and Heine boilers are examples of the water-tube type.

Fire-Tube Boilers. The *Cornish and Lancashire* boilers have the fire tubes of such a diameter that the furnaces may be constructed inside of them. They differ only in the number of cylindrical tubes in which the furnaces are placed, as many as three tubes being placed in the largest sizes (seldom used) of the Lancashire boilers. They are made up to 200-pound steam pressure and possess the following features:

1. High efficiency at moderate rates of combustion
2. Low rate of depreciation
3. Large water space
4. Easily cleaned
5. Large floor space required
6. Cannot be readily forced

The *Galloway boiler* differs from the Lancashire boiler in that there are cross-tubes in the flues.

In the *multitubular boiler*, the number of tubes is greatly increased and their size is diminished. Their heating surface is large and they steam rapidly. They require a separate furnace and are used extensively for power-station work.

Marine boilers require no setting. Among their advantages and disadvantages may be mentioned:

1. Exceedingly small space necessary
2. Radiating surface reduced
3. Good economy
4. Heavy and difficult to repair
5. Unsuitable for bad water
6. Poor circulation of water

The economic boiler is a combination of the Lancashire and multitubular boilers, as is the marine boiler. It is set in brickwork and arranged so that the gases pass under the bottom and along the sides of the boiler as well as through the tubes. It way be compared with other boilers from the following points:

1. Small floor space
2. Less radiating surface than the Lancashire boiler
3. Not easily cleaned
4. Repairs rather expensive
5. Requires considerable draft

Water-Tube Boilers. The chief characteristics of the water-tube boilers, of which there are many types, are

1. Moderate floor space
2. Ability to steam rapidly
3. Good water circulation
4. Adapted to high pressure
5. Easily transported and erected
6. Easily repaired
7. Not easily cleaned
8. Rate of deterioration greater than for Lancashire boiler
9. Small water space, hence variation in pressure with varying demands for steam
10. Expensive setting

Initial Cost. As regards first cost, boilers installed for 150-pound pressure and the same rate of evaporation, will run in the following order: Galloway and Marine, highest first cost, Economic, Lancashire, and Babcock and Wilcox. The increase of cost, with increase of steam pressure, is greatest for the Economic and least for the water-tube type.

Deterioration. Deterioration is less with the Lancashire boiler than with the other types.

Floor Space. The floor space occupied by these various types built for 150 pounds pressure and 7,500 pounds of water, evaporated per hour, is given in Table II.

TABLE II

Boiler Floor Space

Kind of Boiler	Floor Space in sq. ft.
Lancashire	408
Galloway	371
Babcock and Wilcox	200
Marine wet-back	120
Economic	210

Efficiency. The percentage of the heat of the fuel utilized by the boiler is of great importance, but it is difficult to get reliable data in regard to this. Table III* will give some idea of the efficiencies of the different types. The efficiency is more a question of proper proportioning of grate and heating surface and condition of boiler than of the type of boiler. Economizers were not used in any of these tests, but they should always be used with the Lancashire type of boiler.

It is well to select a boiler from 20 to 50 pounds in excess of the pressure to be used, as its life may thus be considerably extended, while, when the boiler is new, the safety valve need not be set so near the normal pressure, and there is less steam wasted by the blowing off of this valve. Again, a few extra pounds of steam may be carried just previous to the time the peak of the load is expected. For pressures exceeding 200 or, possibly, 150 pounds, a water-tube boiler should be selected.

In large stations, it is preferable to make the boiler units of

*From Donkin's "Heat Efficiency of Steam Boilers."

TABLE III
Boiler Efficiencies

Kind of Boiler	No. of Experiments	Mean Efficiency of Two Best Experiments	Lowest Efficiency	Mean Efficiency of All Experiments
Lancashire hand-fired	107	79.5	42.1	62.3
Lancashire machine-fired	40	73.0	51.9	64.2
Cornish hand-fired	25	81.7	53.0	68.0
Babcock and Wilcox hand-fired	49	77.5	50.0	64.9
Marine wet-back hand-fired	6	69.6	62.0	66.0
Marine dry-back hand-fired	24	75.7	64.7	69.2

large capacity, to do away as much as possible with the extra piping and fittings necessary for each unit. Water-tube boilers are best

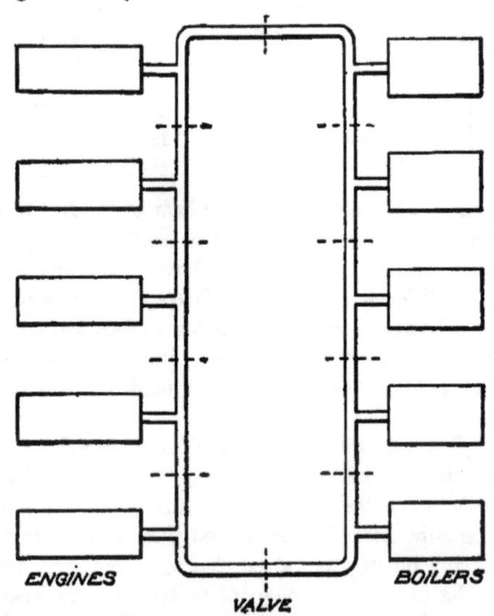

Fig. 2. Diagram of *Ring* System of Piping

adapted for large sizes. These may be constructed for 150-pound pressure, large enough to evaporate 20,000 pounds of water per hour, at an economical rate.

Boilers of the multitubular type or water-tube boilers are used in the majority of power stations in the United States. For stations of moderate size, with medium steam pressures and plenty of space, the return tubular boiler is often employed. For the larger stations and the higher steam pressures, the water-tube boilers are employed. Marine or other special types are used only occasionally where space is limited or where other local conditions govern.

Steam Piping. The piping from the boilers to the engines should be given very careful consideration. Steam should be available at all times and for all engines. Freedom from serious interruptions due to leaks or breaks in the piping is brought about by very careful design and the use of good material in construction.

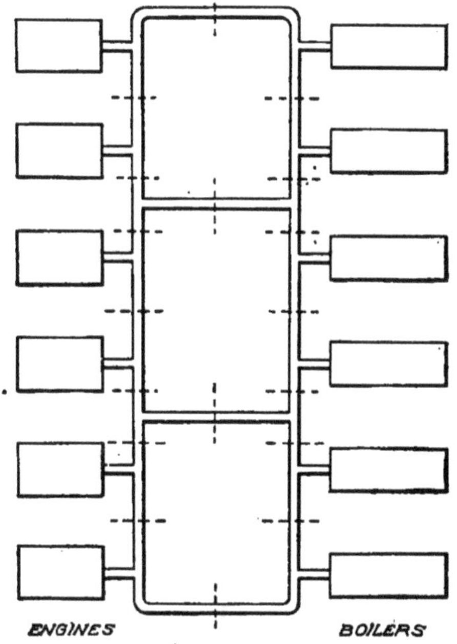

Fig. 3. Diagram of *Ring* System with Cross-Connections

Duplicate piping is used in many instances. Provision must always be made for variations in length of the pipe with variation of temperature. For plants using steam at 150-pound pressure, the variation in the length of steam pipe may be as high as 2.5 inches for 100 feet, and at least 2 inches for 100 feet should always be counted upon.

Arrangement. Fig. 2 shows a simple diagram of the *ring system* of piping. The steam passes from the boiler by two paths to the engine and any section of the piping may be cut out by the closing of two valves. Simple ring systems have the following characteristics:

1. The range, as the main pipe is called, must be of uniform size and large enough to carry all of the steam when generated at its maximum rate.
2. A damaged section may disable one boiler or one engine.
3. Several large valves are required.
4. Provision may readily be made to allow for expansion of pipes.

Cross-connecting the ring system, as shown in Fig. 3, changes these characteristics as follows:

1. Size of pipes and consequent radiating surface is reduced.
2. More valves are needed but they are of smaller size.
3. Less easy to arrange for expansion of the pipes.

If the system is to be duplicated, that is, two complete sets of main pipes and feeders installed, Fig. 4, two schemes are in use:

1. Each system is designed to operate the whole station at maximum load with normal velocity and loss of pressure in the pipes, and only one system is in use at a time. This has the disadvantage that the idle section is liable not to be in good operating condition when needed. Large pipes must be used for each set of mains.
2. The two systems may be made large enough to supply steam at normal loss of pressure when both are used at the same time, while either is made large enough to keep the station running should the other section need repairs. This has the advantages of less expense, and both sections of pipe are normally in use; but it has the disadvantages of more radiating surface to the pipes and consequent condensation for the same capacity for furnishing steam.

Complete interchangeability of units cannot be arranged for if the separate engine units exceed 400 to 500 horse-power. Since engine units can be made larger than boiler units, it becomes necessary to treat several boiler units as a single unit, or battery, these batteries being connected as the single boilers already shown. For still larger plants the steam piping, if arranged to supply any engines from any batteries of boilers, would be of enormous size. If the boilers do not occupy a greater length of floor space than the engines, Fig. 5 shows a good arrangement of units. Any engine can be fed from either of two batteries of boilers and the liability of serious interruptions of service due to steam pipes or boiler trouble is very remote.

In many plants but a single steam range is used, the station depending upon good material, careful construction, and thorough

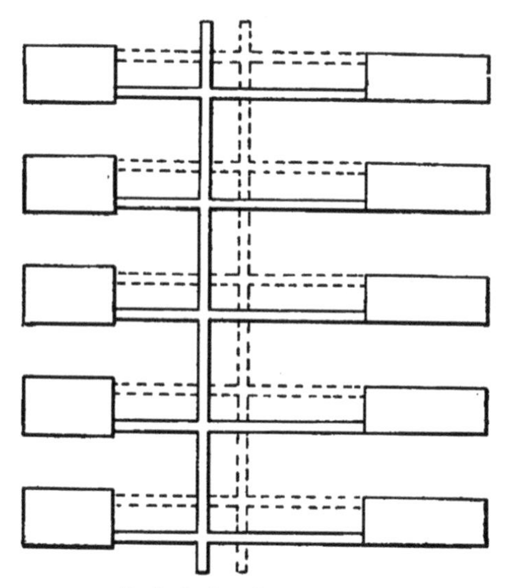

Fig. 4. Duplicate System of Piping

inspection for reliability of service. In the largest steam-turbine stations, the so-called unit system is employed as is explained later under "Station Arrangement."

Material. Steel pipe, lap welded and fastened together by means of flanges, is to be recommended for all steam piping. The flanges may be screwed on the ends of sections and calked so as to render this connection steam tight, though in large sizes it is better to have the flanges welded to the pipes. This latter construction costs no more for large pipes and is much more reliable. All valves and fittings are made in two grades or weights, one for low pressures, and the other for high pressures. The high-pressure fittings should always be used for electrical stations. Gate valves should always

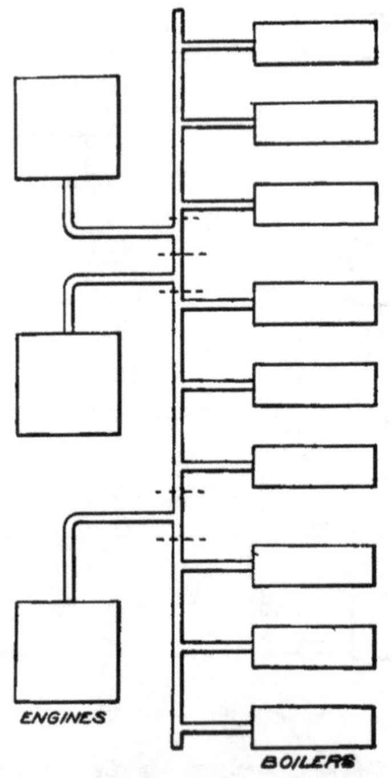

Fig. 5. Arrangement of Boilers and Engines in Very Large Plants

be selected, and, in large sizes, they should be provided with a by-pass.

Asbestos, either alone or with copper rings, vulcanized India rubber, asbestos and India rubber, etc., are used for packing between flanges to render them steam tight. Where there is much expansion, the material selected should be one that possesses considerable

1192- GENERATOR FLOOR.
JAN.-5-1914.

elasticity. Joints for high-pressure systems require much more care than those for low-pressure systems, and the number of joints should be reduced to a minimum by using long sections of pipe.

Fittings. A list of the various fittings required for steam piping, together with their descriptions, is given in books on boilers. One precaution to be taken is to see that such fittings do not become too numerous or complicated, and it is well not to depend too much on automatic fittings. Steam separators should be large enough to serve as a reservoir of steam for the engine and thus equalize, to a certain extent, the velocity of flow of steam in the pipes.

Expansion. In providing for the expansion of pipes due to change of temperature, U bends made of steel pipe and having a radius of curvature not less than six times, and preferably ten times the diameter of the pipe, are preferred. Copper pipes cannot be recommended for high pressures, while slip expansion joints are most undesirable on account of their liability to bind.

Size. The size of steam pipes is determined by the velocity of flow. Probably an average velocity of 60 feet per second would be better than 100 feet per second, though in some cases where space is limited a velocity as high as 150 feet per second has been used.

Loss in Pressure. The loss in pressure in steam pipes may be obtained from the formula

$$p_1 - p_2 = \frac{Q^2 wL}{c^2 d^5}$$

where $p_1 - p_2$ is the loss in pressure in pounds per square inch; Q is the quantity of steam in cubic feet per minute; d is the diameter of pipe in inches; L is the length in feet; w is the weight per cubic feet of steam at pressure p_1 and c is a constant, depending on size of pipe, values of which for the variation in the size of pipe are as follows:

Diameter of pipe	½"	1"	2"	3"	4"	5"	6"	7"	8"	9"	10"
Value of c	36.8	45.3	52.7	56.1	57.8	58.4	59.5	60.1	60.7	61.2	61.8

Diameter of pipe	12"	14"	16"	18"	20"	22"	24"
Value of c	62.1	62.3	62.6	62.7	62.9	63.2	63.2

Mounting. In mounting the steam pipe, it should be fastened rigidly at one point, preferably near the center of a long section, and allowed a slight motion longitudinally at all other supports. Such supports may be provided with rollers to allow for this motion, or the pipe may be suspended from wrought-iron rods which will give a flexible support.

Location. Practice differs in the location of the steam piping, some engineers recommending that it be placed underneath the engine-room floor and others that it be located high above the engine-room floor. In any case it should be made easily accessible, and the valves should be located so that nothing will interfere with their operation. Proper provision must be made for draining the pipes.

Lagging. All piping as well as joints should be carefully covered with a good quality of lagging as the amount of steam condensed in a bare pipe, especially if of any great length, is considerable. In selecting a lagging bear in mind that the covering for steam pipes should be incombustible, should present a smooth surface, should not be damaged easily by vibration or steam, and should have as large a resistance to the passage of heat as possible. It must not be too thick, otherwise the increased radiating surface will counterbalance the resistance to the passage of heat.

The loss of power in steam pipes due to radiation is

$$H = .262\, r\, L\, d$$

where H is loss of power in heat units; d is diameter of pipe in inches; L is the length of pipe in feet; and r is a constant depending on steam pressure and pipe covering, values of which for the variations of these two factors are as follows:

Steam pressure in pounds (absolute)......	40	65	90	115
Values of r for uncovered pipe............	437	555	620	684
Value of r for pipe covered with 2 inches of hair felt..........................	48	58	66	73

Referring to tables in books on boilers, the relative values of different materials used for covering steam pipes may be found.

Superheated Steam. Superheated steam reduces condensation in the engines as well as in the piping, and increases the efficiency of the system. Its use was abandoned for several years, due to difficulties in lubricating and packing the engine cylinders, but by the use of mineral oils and metallic packing, these difficulties have been done away with to a large extent, while steam turbines are especially adapted to the use of superheated steam. The application of heat directly to steam, as is done in the superheater, increases the

TABLE IV

Boiler Efficiencies

Amount of Superheat	Water Evaporated per Pound of Coal	
	Without Superheat	With Superheat
40 degrees F.	7.82	9.99
42 degrees F.	6.42	7.06
55 degrees F.	6.00	7.00
56.5 degrees F.	6.78	8.66
55.2 degrees F.	7.15	8.65

efficiency of the boilers. Table IV shows the increase in boiler efficiency for a certain boiler test, the results being given in pounds of water changed to dry, saturated steam. Tests on various engines show a gain in efficiency as high as 9 per cent with a superheat of 80° to 100° F., while special tests in some cases show even a greater gain.

Superheaters are very simple, consisting of tubular boilers containing steam instead of water, and either located so as to utilize the heat of the gases, the same as economizers, or separately fired. They should be arranged so that they may be readily cut out of service, if necessary, and provision must be made for either flooding them or turning the hot gases into a by-pass, as the tubes would be injured by the heat if they contained neither water nor steam. Superheaters may be mounted in the furnace of the regular boiler setting or they may have furnaces of their own and be separately fired. For electrical stations using superheated steam the former type is usually employed and it has proved very satisfactory for moderate degrees of superheat.

Feed Water. All water available for the feeding of boilers contain some impurities, among the most important of which as regards boilers are soluble salts of calcium and magnesium. Bicarbonates of the alkaline earths cause precipitations on the interior of boilers, forming *scale*. Sulphate of lime is also deposited by concentration under pressure. Scale, when formed, not only decreases the efficiency of the boiler but also causes deterioration, for if sufficiently thick, the diminished conducting power of the boiler allows the tubes or plates to be overheated and to crack or burst. Again, the scale may keep the water from contact with sections of the heated plates for some time and then, giving way, large volumes of steam are generated very quickly, and an explosion may result.

Some processes to prevent the formation of scale are used, which affect the water after it enters the boilers, but they are not to be recommended, and any treatment the water receives should affect it previous to its being fed to the boilers. Carbonates and a small quantity of sulphate of lime may be removed by heating in a separate vessel. Large quantities of sulphate of lime must be precipitated chemically.

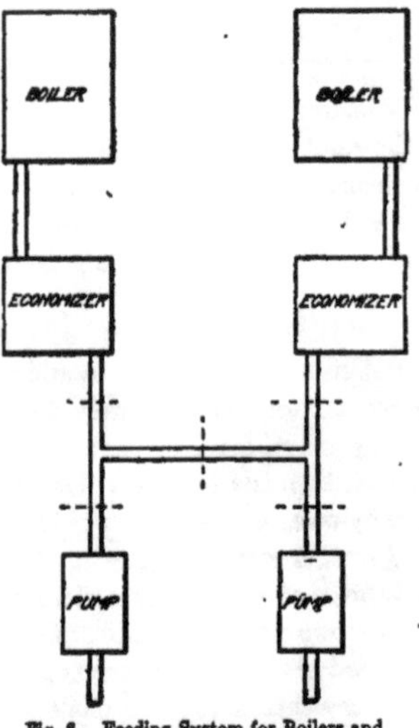

Fig. 6. Feeding System for Boilers and Pumps

Sediment must be removed by allowing the water to settle. Vegetable matters are sometimes present, which cause a film to be deposited. Certain gases, in solution—such as oxygen, nitrogen, etc.—cause pitting of the boiler. This effect is neutralized by the addition of chemicals. Oil from the engine cylinder is particularly destructive to boilers and when present in the condensed steam must be carefully removed.

Feeding Appliances. Both feed pumps and injectors are used for feeding the water to the boilers. *Feed pumps* may be either steam- or motor-driven. Steam-driven pumps are very inefficient, but they are simple and the speed is easily controlled. Motor-driven pumps are more efficient and neater, but more expensive and more difficult to regulate efficiently over a wide range of speed. Direct-acting pumps may have feed-water heaters attached to them, thus increasing the efficiency of the apparatus as a whole. The supply of electrical energy must be constant if motor-driven pumps are to be used.

TABLE V

Rate of Flow of Water, in Feet per Minute, Through Pipes of Various Sizes, for Varying Quantities of Flow

Gallons per Min.	¾ in.	1 in.	1¼ in.	1½ in.	2 in.	2½ in.	3 in.	4 in.
5	218	122½	78½	54½	30½	19½	13½	7¾
10	436	245	157	109	61	38	27	15¼
15	653	367½	235½	163½	91½	58½	40½	23
20	872	490	314	218	122	78	54	30¾
25	1090	612½	392½	272½	152½	97½	67½	38¼
30		735	451	327	183	117	81	46
35		857½	549½	381½	213½	136½	94½	53¾
40		980	628	436	244	156	108	61¼
45		1102½	706½	490½	274½	175½	121½	69
50			785	545	305	195	135	76¾
75			1177½	817½	457½	292½	202½	115
100				1090	610	380	270	153¼
125					762½	487½	337½	191¾
160					915	585	405	230
175					1067½	682½	472½	268¼
200					1220	780	540	306¾

Feed pipes must be arranged so as to reduce the risk of failure to a minimum, and for this reason they are almost always duplicated. More than one water supply is also recommended if there is the slightest danger of interruption on this account. One common arrangement of feed-water apparatus is to install a few large pumps supplying either of two mains from which the boiler connections are taken. This is a complicated and costly system of piping. A scheme for feeding two boilers where each pump is capable of supplying both boilers is shown in Fig. 6. Pipes should be ample in cross-section, and, in long lengths, allowance must be made for expansion. Cast iron or cast steel is the material used for their construction; the joints being made by means of flanges fitted with rubber gaskets.

The rate of flow of water in feet per minute through pipes of various sizes is given in Table V. A flow of 10 gallons per minute for each 100 h. p. of boiler equipment should be allowed without causing an excessive velocity of flow in the pipes—400 to 600 feet per minute represents a fair velocity.

Boiler Setting. The economical use of coal depends, to a large extent, on the setting of the boiler and proper dimensions of the furnaces. Internally-fired boilers require support only, while the setting of externally-fired boilers requires provision for the furnaces. Common brick, together with fire brick for the lining of portions exposed to the hot gases, are used almost invariably for boiler settings. It is customary to set the boiler units up in batteries of two, using a 20-inch wall at the sides and a 12-inch wall between the two boilers. The instructions for settings furnished by the manufacturers should be carefully followed out as they are based on conditions which give the best results in the operation of their boilers.

Draft. The best ratio of heating to grate surface for boiler plants depends upon the kind of fuel used and the draft employed. Based on a draft of 0.5 inch of water, the following values are given for different grades of fuel:

Pocahontas, W. Va., 45; Youghiogheny, Pa., 48; Hocking Valley, O., 45; Big Muddy, Ill., 50; Lackawanna, Pa., No. 1 buckwheat, 32. The first of these coals is semi-bituminous, the Lackawanna coal is anthracite, and the other coals are bituminous.

Natural Draft. Natural draft is the most commonly used and is the most satisfactory under ordinary circumstances. In determining the size of the chimney necessary to furnish this draft, the following formula is given by Kent:

$$A = \frac{.06\,F}{\sqrt{h}} \qquad \text{or } h = \left(\frac{.06\,F}{A}\right)^2$$

where A = area of chimney in sq. ft.; h = height of chimney in ft.; and F = pounds of coal per hour.

The height of chimney should be assumed and the area calculated, remembering that it is better to have the chimney too large than too small.

The chimney may be either of brick or iron, the latter having a less first cost but requiring repairs at frequent intervals. General rules for the design of a brick chimney may be given as follows:

The external diameter of the base should not be less than $\frac{1}{10}$ of the height.
Foundations must be of the best.
Interiors should be of uniform section and lined with fire brick.
An air space must exist between the lining and the chimney proper.

The exterior should have a taper of from $\frac{1}{16}$ to $\frac{1}{4}$ inch to the foot. Flues should be arranged symmetrically.

Fig. 7 shows the construction of a brick chimney of good design, this chimney being used with boilers furnishing engines which develop 14,000 h. p.

Mechanical Draft. Mechanical draft is a term which may be used to embrace both forced and induced draft. The first cost of mechanical-draft systems is less than that of a chimney, but the operation and repair are much more expensive and there is always the risk of break-down. Artificial draft has the advantage that it can be varied within large limits and it can be increased to any desired extent, thus allowing the use of low grades of coal.

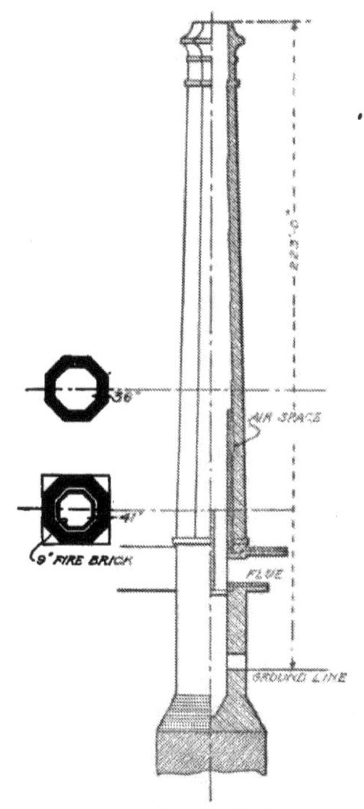

Fig. 7. Good Design of Brick Chimney

Firing of Boilers. Coal is used for fuel to a greater extent than any other material, though oil, gas, wood, etc., are used in some localities. Local conditions, such as availability, cost, etc., should determine the material to be used; no general rules can be given. From data regarding the relative heating values of different fuels we find: that 1 pound of petroleum, about $\frac{1}{7}$ of a gallon, is equivalent, when used with boilers, to 1.8 pounds of coal and there is less deterioration of the furnace with oil; that $7\frac{1}{2}$ to 12 cubic feet of natural gas are required as the equivalent of 1 pound of coal, depending on the quality of the gas; that $2\frac{1}{2}$ pounds of dry wood is assumed as the equivalent of 1 pound of coal.

Stoking. When coal is used, it requires stoking and this may be accomplished either by hand or by means of mechanical stokers, many forms of which are available. Mechanical stoking has the advantage over hand stoking in that the fuel may be fed to the furnace more uniformly, thus avoiding the subjection of the fires and boilers to sudden blasts of cold air as is the case when the fire doors are opened; in that a poorer grade of coal may be burned, if necessary; and in that the trouble due to smoke is much reduced. It may be said that mechanical stokers are used almost universally in the more important electrical plants. Economic use of fuel requires great care in firing, especially if it is done by hand.

Where gas is used, the firing may be made nearly automatic, and the same is true of oil firing, though the latter requires more complicated burners, as it is necessary that the oil be vaporized.

In large stations, operated continuously, it is desirable that, as far as possible, all coal and ashes be handled by machinery, though the difference in cost of operation should be carefully considered before installing extensive coal-handling machinery. Machinery for automatically handling the coal will cost from $7.50 to $10 per horse-power rating of boilers for installation, while the ash-handling machinery will cost from $1.50 to $3 per horse-power.

The coal-handling devices usually consist of chain-operated conveyors which hoist the coal from railway cars, barges, etc., to overhead bins from which it may be fed to the stokers. The ashes may be handled in a similar manner, by means of scraper conveyors, or small cars may be used. Either steam or electricity may be used for driving this auxiliary apparatus.

It is always desirable that there be generous provision for the storage of fuel sufficient to maintain operations of the plant over a temporary failure of supply.

STEAM ENGINES

The choice of steam prime movers is one which is governed by a number of conditions which can be treated but briefly here. The first of these conditions relates to the speed of the engine to be used. There is considerable difference of opinion in regard to this as both high- and low-speed plants are in operation and are giving good satisfaction. Slow-speed engines have a higher first cost and a higher economy. Probably in sizes up to 250 kw., the generator should be driven by high-speed engines; from 250 to 500 kw., the selection of either type will give satisfaction; above 500 indicated horse-power, the slow-speed type is to be recommended. Drop valves cannot be used with satisfaction for speeds above about 100 revolutions per minute, hence high-speed engines must use direct-driven valve gears, usually governed by shaft governors. Corliss valves are used on nearly all slow-speed engines.

The steam pressure used should be at least 125 pounds per square inch at the throttle and a pressure as high as 150 to 160 pounds is to be preferred.

Close regulation and uniform angular velocity are required for driving generators, especially alternators which are to operate in parallel. This means sensitive and active governors, carefully designed flywheels, and proper arrangement of cranks when more than one is used.

High-speed engines should not have a speed change greater than 1½ per cent from no load to full load, but for prime movers used for driving large alternators operated in multiple, a speed change as great as 4 or 5 per cent may be desirable. The variation in angular velocity, where alternators are to be operated in parallel, should be within such limits that at no time will the rotating part be more than $\frac{1}{60}$ of the pitch angle of two poles from the position it would occupy if the angular velocity were uniform at its mean value.

For large engine-driven plants or plants of moderate size, compound condensing engines are almost universally installed. The advantage of these engines in increased economy are in part counterbalanced by higher first cost and increased complications, together with the pumps and added water supply necessary for the condensers. The approximate saving in amount of steam is shown in Table VI, which applies to a 500 horse-power unit.

Triple expansion engines are seldom used for driving electrical machinery, as their advantages under variable loads are doubtful. Compound engines may be tandem or cross-compound and either horizontal or vertical. The use of cross-compound engines tends to produc^ uniform angular velocity, but the cylinder should be so proportioned that the amount of work done by each is nearly equal.

TABLE VI

Engine	Pounds of Steam per H. P. Hour
Simple non-condensing	30
Simple condensing	22
Compound non-condensing	24
Compound condensing	16

A cylinder ratio of about 3½ to 1 will approximate average conditions. Either vertical or horizontal engines may be installed, each having its own peculiar advantages. Vertical engines require less floor space, while horizontal engines have a better arrangement of parts. Either type should be constructed with heavy parts and erected on solid foundations.

Engines should preferably be direct-connected, but this is not always feasible, and gearing, belt, or rope drives must be resorted to. Countershafts, belt or rope driven, arranged with pulleys and belts for the different generators, and with suitable clutches, are largely used in small stations. They consume considerable power and the bearings require attention.

Careful attention must be given to the lubrication of all running parts, and extensive oil systems are necessary in large plants. In such systems a continuous circulation of oil over the bearings and through the engine cylinders is maintained by means of oil pumps. After passing through the bearings, the machine oil goes to a properly arranged oil filter where it is cleaned and then pumped to the bearings again. A similar process is used in cylinder lubrication, the oil being collected from the exhaust steam, and only enough new oil is added to make up for the slight amount lost. The latter system is not installed as frequently as the continuous system for bearings.

STEAM TURBINES

Advantages. The steam turbine is now very extensively used as a prime mover for generators in power stations on account of its many advantages, some of which may be stated as follows:

1. High steam economy at all loads.
2. High steam economy with rapidly fluctuating loads.

3. Small floor space per kw. capacity, reducing to a minimum the cost of real estate and buildings.
4. Uniform angular velocity, thus facilitating the parallel operation of alternators.
5. Simplicity in operation and low expense for attendance.
6. Freedom from vibration, hence low cost for foundations.
7. Steam economy is not appreciably impaired by wear or lack of adjustment in long service.
8. Adaptability to high steam pressures and high superheat without difficulty in operation and with consequent improvement in economy.
9. Condensed steam is kept entirely free from oil and can be returned to the boilers without passing through an oil separator.

Types. The detailed descriptions of the different types of steam turbines are given in books devoted to steam engines and turbines and only a small amount of space can be devoted to them here. The first classification of steam turbines is into the impulse type and the reaction type of turbine. In the impulse type the steam is expanded in passing through suitable nozzles and does useful work in moving the blades of the rotating part by virtue of its kinetic energy. In the reaction type the steam is only partially expanded before it comes into contact with the blades and much of the work on the moving blades is accomplished by the further expansion of the steam and the reaction of the steam as it leaves the blades. Of the impulse type the *DeLaval* and the *Curtis* turbines are well-known makes. The DeLaval turbine is built in small and moderate sizes only and is of the single-stage type. The Curtis turbine is built in all sizes up to the very largest and is of the multi-stage type. The Curtis turbine may be briefly described as follows:

The Curtis turbine is divided into sections or stages, each stage containing one or more sets of stationary vanes and revolving buckets. These vanes and buckets are supplied with steam which passes through suitable nozzles to give it the proper expansion and velocity as it issues from the nozzles. By dividing the work into stages, the nozzle velocity of the steam is kept down to a moderate value in each stage and the energy of the steam is effectively given up to the rotating part without excessively high speeds. Fig. 8

shows the arrangement of nozzles, buckets, and stationary blades or guiding vanes for two stages. A complete turbine of the vertical type and of 5,000 kw. capacity is shown in Fig. 9. Governing is accomplished by automatically opening or closing some of the nozzles,

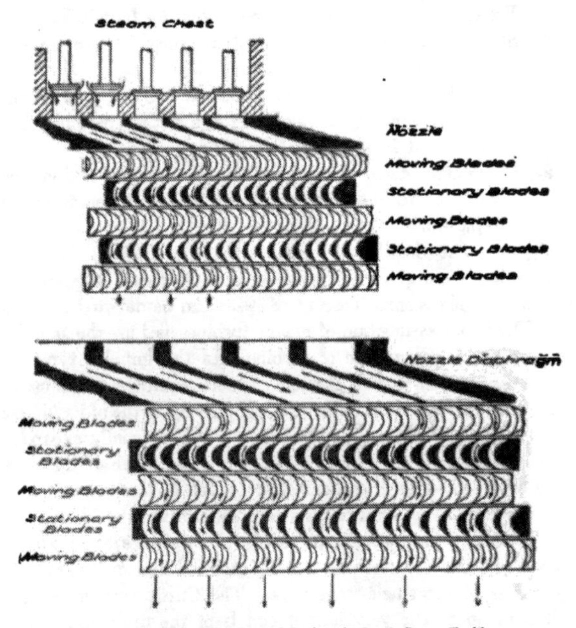

Fig. 8. Diagram of Nozzles and Buckets in Curtis Steam Turbine

and on overloads the steam may be automatically led directly into the second stage of the turbine. The step bearing which supports the weight of the rotating part may be lubricated by either oil or water under high pressure, this pressure being made great enough to support the weight of the moving element on a thin film of the lubricant. Only a vertical type of the Curtis turbine is shown here but it is also manufactured in the horizontal form.

Of the reaction turbines the *Parson's* type is the most prominent one. It is manufactured in the United States by the Westinghouse Machine Company and the Allis-Chalmers Company. An elementary drawing of the cross-section of the Allis-Chalmers turbine is shown in Fig. 10. Steam enters this turbine at *C* through the governing valve *D*, passes through the opening *E*, and thence expands in its passages through the series of revolving and stationary blades

Fig. 9. Turbo-Alternator of 5,000 kw. Capacity

in the three stages *H*, *J*, and *K*. The steam pressure is balanced by means of a series of disks or balance pistons shown at *L*, *M*, and *N*. The valve shown at *V* is automatically opened on overload, thus admitting steam directly into stage *J*.

The steam economy of the turbine increases with increase in vacuum approximately as follows: For every increase in vacuum of one inch between 23 inches and 28 inches the increase in economy is 3 per cent for 100-kw. units, 4 per cent for 400-kw. units, and 5 per cent for 1,000-kw. units. This is a greater improvement than can be obtained with steam engines under corresponding conditions

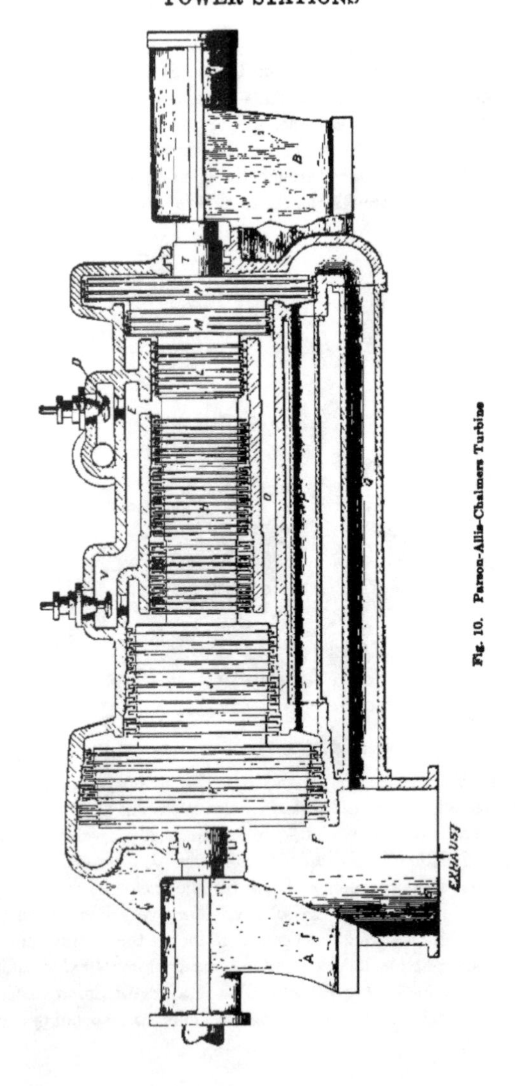

Fig. 10. Parson-Allis-Chalmers Turbine

and the exhaust-steam or low-pressure turbine is being introduced to work in conjunction with the reciprocating steam engine, the steam expanding down to about atmospheric pressure in the engine and continuing down to a high vacuum through the low-pressure turbine. A receiver may be introduced between the engine and the turbine. A higher steam economy is claimed for such a combination than could be secured by either engine or turbine alone.

HYDRAULIC PLANTS

Because of the relative ease with which electrical energy may be transmitted long distances, it has become quite common to locate large power stations where there is abundant water power, and to transmit the energy thus generated to localities where it is needed. This type of plant has been developed to the greatest extent in the western part of the United States, where in some cases the transmission lines are very extensive. The power houses now completed, or in the course of erection at Niagara Falls, are examples of the enormous size such stations may assume.

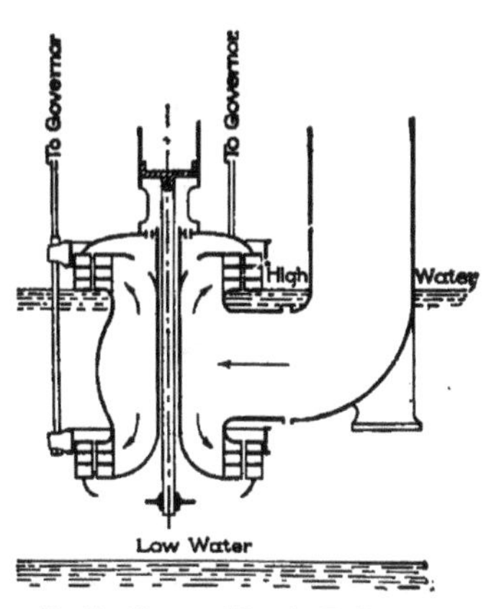

Fig. 11. Diagram of Reaction Turbines

Before deciding to utilize water power for driving the machinery in central stations, the following points should be noted:

1. The amount of water power available.
2. The possible demand for power.
3. Cost of developing this power as compared with cost of plants using other sources of power.
4. Cost of operation compared with other plants and extent of transmission lines.

Hydraulic plants are often much more expensive than steam

plants, but the first cost is more than made up by the saving in operating expenses.

Methods for the development of water powers vary with the nature and the amount of the water supply, and they may be studied best by considering plants which are in successful operation, each one of which has been a special problem in itself. A full description of such plants would be too extensive to be incorporated here, but they can be found in the various technical journals.

Water Turbines. Water turbines used for driving generators are of two general classes, *reaction* turbines and *impulse* turbines.

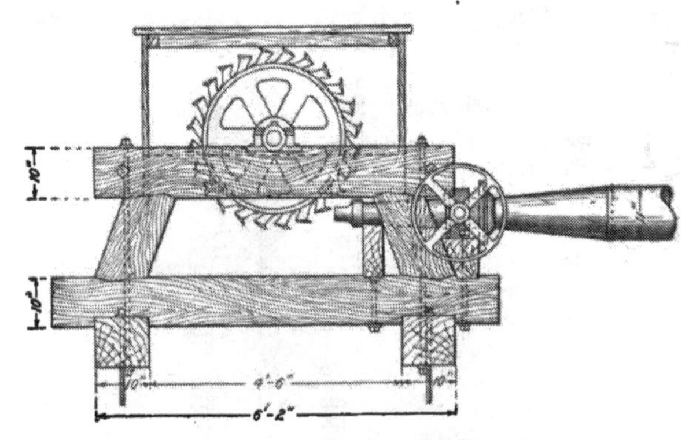

Fig. 12. Pelton Type of Impulse Turbine

Reaction turbines may be subdivided into *parallel-flow, outward-flow,* and *inward-flow* types. Parallel-flow turbines are suited for low falls, not exceeding 30 feet. Their efficiency is from 70 to 72 per cent. Outward-flow and inward-flow turbines give an efficiency from 79 to 88 per cent. Impulse turbines are suitable for very high falls and should be used from heads exceeding, say, 100 feet, though it is difficult to say at what head the reaction turbine would give place to the impulse wheel, as reaction turbines are giving good satisfaction on heads in the neighborhood of 200 feet, while impulse wheels are operated with falls of but 80 feet. A reaction wheel is shown in Fig. 11, and the Pelton wheel, one of the best known types

6600- AND 3500-VOLT VERTICAL SWITCHBOARD OF THE KYOTO ELECTRIC COMPANY OF JAPAN
Courtesy of General Electric Company

TABLE VII

Pressure of Water

Feet Head	Pressure Pounds per Sq. In.	Feet Head	Pressure Pounds per Sq. In.	Feet Head	Pressure Pounds per Sq. In.	Feet Head	Pressure Pounds per Sq. In.
10	4.33	105	45.48	200	86.63	295	127.78
15	6.49	110	47.64	205	88.80	300	129.95
20	8.66	115	49.81	210	90.96	310	134.28
25	10.82	120	51.98	215	93.13	320	138.62
30	12.99	125	54.15	220	95.30	330	142.95
35	15.16	130	56.31	225	97.46	340	147.28
40	17.32	135	58.48	230	99.63	350	151.61
45	19.49	140	60.64	235	101.79	360	155.94
50	21.65	145	62.81	240	103.90	370	160.27
55	23.82	150	64.97	245	106.13	380	164.61
60	25.99	155	67.14	250	108.29	390	168.94
65	28.15	160	69.31	255	110.46	400	173.27
70	30.32	165	71.47	260	112.62	500	216.58
75	32.48	170	73.64	265	114.79	600	259.90
80	34.65	175	75.80	270	116.96	700	303.22
85	36.82	180	77.97	275	119.12	800	346.54
90	38.98	185	80.14	280	121.29	900	389.86
95	41.15	190	82.30	285	123.45	1000	433.18
100	43.31	195	84.47	290	125.62		

of impulse wheels, is shown in Fig. 12. An efficiency as high as 86 per cent is claimed for the impulse wheel under favorable conditions. The fore bay leading to the flume should be made of such size that the velocity of water does not exceed 1½ feet per second; and it should be free from abrupt turns. The same applies to the tailrace. The velocity of water in wooden flumes should not exceed 7 to 8 feet per second. Riveted steel pipe is used for the penstocks and for carrying water from considerable distances under high heads. In some locations it is buried, in others it is simply placed on the ground. Wooden-stave pipe is used to a large extent when the heads do not much exceed 200 feet. In Table VII is given the pressure of water in pounds per square inch at different heads, while in Table VIII is given considerable data relating to riveted-steel hydraulic pipe. Governors of the usual types are required to keep the speed of the turbine constant under change of load and change of head.

TABLE VIII

Riveted Hydraulic Pipe

Diam. of Pipe in Inches	Area of Pipe in Square Inches	Thickness of Iron by Wire Gauge	Head in Feet the Pipe will Safely Stand	Cu. Ft. Water Pipe will Convey per Min. at Vel. 3 Ft. per Sec.	Weight per Lineal Foot in Pounds
3	7	18	400	9	2
4	12	18	350	16	2¼
4	12	16	525	16	3
5	20	18	325	25	3½
5	20	16	500	25	4¼
5	20	14	675	25	5
6	28	18	296	36	4¼
6	28	16	487	36	5¾
6	28	14	743	36	7½
7	38	18	254	50	5¼
7	38	16	419	50	6¾
7	38	14	640	50	8½
8	50	16	367	63	7½
8	50	14	560	63	9½
8	50	12	854	63	13
9	63	16	327	80	8½
9	63	14	499	80	10¾
9	63	12	761	80	14¼
10	78	16	295	100	9¼
10	78	14	450	100	11¾
10	78	12	687	100	15¾
10	78	11	754	100	17½
10	78	10	900	100	19½
11	95	16	269	120	9¾
11	95	14	412	120	13
11	95	12	626	120	17¼
11	95	11	687	120	18¾
11	95	10	820	120	21
12	113	16	246	142	11¼
12	113	14	377	142	14
12	113	12	574	142	18½
12	113	11	630	142	19¾
12	113	10	753	142	22¾
13	132	16	228	170	12
13	132	14	348	170	15
13	132	12	530	170	20
13	132	11	583	170	22
13	132	10	696	170	24½
14	153	16	211	200	13
14	153	14	324	200	16
14	153	12	494	200	21½
14	153	11	543	200	23½
14	153	10	648	200	26
15	176	16	197	225	13¾
15	176	14	302	225	17
15	176	12	460	225	23
15	176	11	507	225	24½
15	176	10	606	225	28
16	201	16	185	255	14½
16	201	14	283	255	17¼
16	201	12	432	255	24¼
16	201	11	474	255	26½
16	201	10	567	255	29½

Riveted Hydraulic Pipe
(Continued)

Diam. of Pipe in Inches	Area of Pipe in Square Inches	Thickness of Iron by Wire Gauge	Head in Feet the Pipe will Safely Stand	Cu. Ft. Water Pipe will Convey per Min. at Vel. 3 Ft. per Sec.	Weight per Lineal Foot in Pounds
18	254	16	165	320	16½
18	254	14	252	320	20½
18	254	12	385	320	27¼
18	254	11	424	320	30
18	254	10	505	320	34
20	314	16	148	400	18
20	314	14	227	400	22½
20	314	12	346	400	30
20	314	11	380	400	32½
20	314	10	456	400	36½
22	380	16	135	480	20
22	380	14	206	480	24¾
22	380	12	316	480	32¾
22	380	11	347	480	35¾
22	380	10	415	480	40
24	452	14	188	570	27¼
24	452	12	290	570	35½
24	452	11	318	570	39
24	452	10	379	570	43½
24	452	8	466	570	53
26	530	14	175	670	29¼
26	530	12	267	670	38½
26	530	11	294	670	42
26	530	10	352	670	37
26	530	8	432	670	57¼
28	615	14	102	775	31¼
28	615	12	247	775	41¼
28	615	11	273	775	45
28	615	10	327	775	50¼
28	615	8	400	775	61¼
30	706	12	231	890	44
30	706	11	254	890	48
30	706	10	304	890	54
30	706	8	375	890	65
30	706	7	425	890	74
36	1017	11	141	1300	58
36	1017	10	155	1300	67
36	1017	8	192	1300	78
36	1017	7	210	1300	88
40	1256	10	141	1600	71
40	1256	8	174	1600	86
40	1256	7	189	1600	97
40	1256	6	213	1600	108
40	1256	4	250	1600	126
42	1385	10	135	1760	74½
42	1385	8	165	1760	91
42	1385	7	180	1760	102
42	1385	6	210	1760	114
42	1385	4	240	1760	133
42	1385	¼	270	1760	137
42	1385	3	300	1760	145
42	1385	5/16	321	1760	177
42	1385	⅜	363	1760	216

TABLE IX

Horse-Power per Cubic Foot of Water per Minute for Different Heads

Heads in Feet	Horse-Power	Heads in Feet	Horse-Power	Heads in Feet	Horse-Power	Heads in Feet	Horse-Power
1	.0016098	170	.273666	330	.531234	490	.788802
20	.032196	180	.289764	340	.547332	500	.804900
30	.048294	190	.305862	350	.563430	520	.837096
40	.064392	200	.321960	360	.579528	540	.869292
50	.080490	210	.338058	370	.595626	560	.901488
60	.096588	220	.354156	380	.611724	580	.933684
70	.112686	230	.370254	390	.627822	600	.965880
80	.128784	240	.386352	400	.643920	650	1.046370
90	.144892	250	.402450	410	.660018	700	1.126860
100	.160980	260	.418548	420	.676116	750	1.207350
110	.177078	270	.434646	430	.692214	800	1.287840
120	.193176	280	.450744	440	.708312	900	1.448820
130	.209274	290	.466842	450	.724410	1000	1.609800
140	.225372	300	.482940	460	.740508	1100	1.770780
150	.241470	310	.499038	470	.756606		
160	.257568	320	.515136	480	.772704		

GAS PLANT

The gas engine using natural gas, producer gas, blast furnace gas, or even illuminating gas in some instances, is being used to a considerable extent as a prime mover for electric generators. The advantages claimed for the gas engine are:

1. Minimum fuel and heat consumption.
2. Low cost of operation and maintenance.
3. Simplification of equipment and small number of auxiliaries.
4. No heat lost due to radiation when engines are idle.
5. Quick starting.
6. Extensions may be easily made.
7. High pressures are limited to the engine cylinders.

As disadvantages of the gas engine may be mentioned the large floor space required; small overload capacity; and the heavy and expensive foundations necessary.

Fig. 13 shows the efficiency and amount of gas consumed by a 500-h. p. engine, Pittsburg natural gas being used.

The only auxiliaries needed where natural gas is employed are the igniter generators and the air compressors—with a pump for the jacket water in some cases—which may be driven by either a motor or a separate gas engine. The jacket water may be utilized for heating purposes in many plants. Cooling towers may be installed where water is scarce.

Parallel operation of alternators when direct-driven by gas engines has been successful, a spring coupling being used between

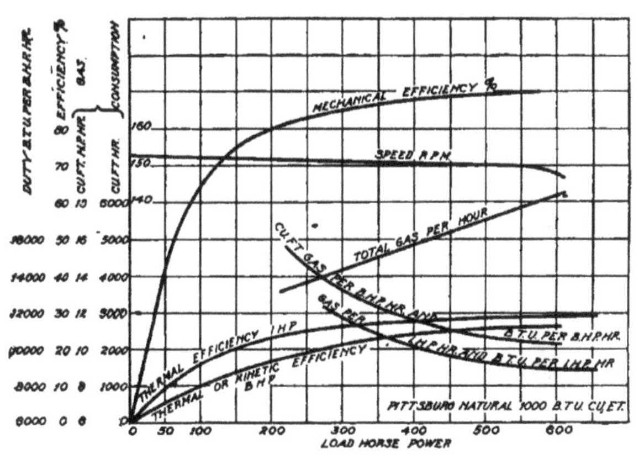

Fig. 13. Efficiency Curves of a 500-H. P. Gas Engine

the engines and the generators in some cases to absorb the variation in angular velocity.

The overload capacity of gas engines depends upon the manner of rating. The ultimate capacity is reached when the engine is using a full charge of the best mixture of gas and air at each power stroke. Many manufacturers rate their engines at 10 per cent below the maximum capacity, thus allowing for a limited amount of overload. The gas consumption of gas engines is relatively high at loads less than 50 per cent of normal; hence, it is desirable that the load be fairly constant and at some value between 50 and 100 per cent of the rating of the machine. H. G. Stott has proposed that the gas engine be combined with the steam turbine in some electrical plants, since the turbine can carry heavy overloads and is fairly economical on all loads. In such a plant the steam turbine would carry the fluctuations, and arrangements would be made so that the gas engine would carry a nearly constant load.

Gas-producers for gas engines are of two types: the *suction producer*, used for small plants and employing high-grade fuels; and the *pressure producer*, used for the larger units and manufactured for all grades of fuel.

The fact that no losses occur, due to heat radiation when the machines are not running, and the lack of losses in piping, add greatly to the plant efficiency. If producer gas or blast-furnace gas is used, a larger engine must be installed to give the same power than when natural or ordinary coal gas is used. Electric stations are often combined with gas works, and gas engines can be installed in such stations to particular advantage in many cases.

In addition to the gas engine, other forms of internal combustion engines, such as oil engines and gasoline engines, are being used to a limited extent in small stations.

ELECTRIC PLANT

GENERATORS

The first thing to be considered in the electric plant is the generators, after which the auxiliary apparatus in the way of exciters, controlling switches, safety devices, etc., will be taken up. A general rule which, by the way, applies to almost all machinery for power stations, is to select apparatus which is considered as "standard" by the manufacturing companies. This rule should be followed for two reasons: *First*, reliable companies employ men who may be considered as experts in the design of their machines, and their best designs are the ones which are standardized. *Second*, standard apparatus is from 15 to 25 per cent cheaper than semi-standard or special work, owing to larger production, and it can be furnished on much shorter notice. Again, repair parts are more cheaply and readily obtained.

Specifications should call for performance, and details should be left, to a very large extent, to the manufacturers. Following are some of the matters which may be incorporated in the specifications for generators:

1. Type and general characteristics.
2. Capacity and overload with heating limits.
3. Commercial efficiency at various loads.
4. Excitation.
5. Speed and regulation.
6. Mechanical features.

Types. The type of machine will be determined by the system selected. Generators may be direct-current or alternating-current—single or polyphase—or as in some plants now in operation, they may be double-current. The voltage, compounding, frequency, etc., should be stated. Direct-current machines are seldom wound for a voltage above 600, but alternating-current generators may be purchased which will give as high as 15,000 volts at the terminals. As a rule it is well not to use an extremely high voltage for the generators themselves, but to use step-up transformers in case a very high line voltage is necessary. Up to about 7,000 volts, generators may be safely used directly on the line. Above this, local conditions will decide whether to connect the machine directly to the line or to step up the voltage. Machines wound for high potential are more expensive for the same capacity and efficiency, but the cost of step-up transformers and the losses in the same are saved by using such machines, so that there is a slight gain in efficiency which may be utilized in better regulation of the system, or in lighter construction of the line. On the other hand, lightning troubles are liable to be aggravated when transformers are not used, as the transformers act as additional protection to the machines, and if the transformers are injured they may be more readily repaired or replaced.

The following voltages are considered standard: Direct-current generators 125, 250, 550-600. Alternating-current systems, high pressure, 2,200, 6,600, 11,000, 22,000, 33,000, 44,000, 66,000, 88,000, and 110,000. The generators, when used with transformers, should be capable of giving a no-load voltage 10 per cent in excess of these figures. Twenty-five and 60 cycles are considered as standard frequencies, the former being more desirable for railway work and the latter for lighting purposes.

Capacity. The size of machines to be chosen has been briefly considered. Alternators are rated for non-inductive load or a power

TABLE X

Average Maximum Efficiencies

Kw.	Per Cent	Kw.	Per Cent
5	85	150	93
10	88	200	94
25	90	500	95
50	92	1000	96

factor of unity unless a different power factor is distinctly stated. Aside from the overload capacity to be counted upon as reserve, the Standardization Report of the American Institute of Electrical Engineers recommends the following for the heating limits and overload capacity of generators:

MAXIMUM VALUES OF TEMPERATURE ELEVATION

Field and armature, by resistance,	50° C.
Commutator and collector rings and brushes, by thermometer,	55° C.
Bearings and other parts of machine, by thermometer,	40° C.

Overload capacity should be 25 per cent for two hours, with a temperature rise not to exceed 15 degrees above full load values, the machine to be at constant temperature reached under normal load, before the overload is applied. A momentary overload of 50 per cent should be permissible without excessive sparking or injury. Some companies recommend an overload capacity of 50 per cent for two hours when the machines are to be used for railway purposes. The above temperature increases are based upon a room-temperature of 25° C.

Efficiency. As a rule, generators should have a high efficiency over a considerable range of load, although much depends upon the nature of the load. It is always desirable that maximum efficiency be as high as is compatible with economic investment.

Table X gives reasonable efficiencies which may be expected for generating apparatus. In order to arrive at what may be considered the best maximum efficiency to be chosen, the cost of power generation must be known, or estimated, and the fixed charges on capital invested must also be a known quantity. From the cost of power, the saving on each per cent increase in efficiency can be determined, and this should be compared with the charges on the

TABLE XI

Exciters for Single-Phase Alternating-Current Generators

60 Cycles

Alternator Classificaton			Exciter Classification		
Poles	Kw.	Speed	Poles	Kw.	Speed
8	60	900	2	1.5	1,900
8	90	900	2	1.5	1,900
8	120	900	2	1.5	1.900
12	180	600	2	2.5	1,900
16	300	450	2	4.5	1,800

additional investment necessary to secure this increased efficiency. A certain point will be found where the sum of the two will be a minimum.

If a generator is to be run for a considerable time at light loads, one with low "no-load" losses should be chosen. These losses are not rigidly fixed but they vary slightly with change of load. It is the same question of "all-day efficiency" which is treated, in the case of transformers, in "Power Transmission." Under no-load losses may be considered, in shunt-wound generators, friction losses, core losses, and shunt-field losses. I^2R losses in the series field, in the armature, and in the brushes, vary as the square of the load.

Excitation. Dynamos, if for direct current, may be self-excited, shunt-wound, compound-wound, or separately excited. Separate excitation is not recommended for these machines. Alternators require separate excitation, though they may be compounded by using a portion of the armature current when rectified by a commutator. Automatic regulation of voltage is always desirable, hence, the general use of compound-wound machines for direct currents. Many alternators using rectified currents in series fields for keeping the voltage nearly constant are in service in small plants, as well as several of the so-called "compensated" alternators, arranged with special devices which maintain the same compounding with different power factors. The latter machine gives good satisfaction if properly cared for, but an automatic regulator, governed by the generator voltage and current, which acts directly on the exciter field, is taking its place. This regulator, known as the Tirrill regulator, is described under "Power Transmission." The capacity of the exciters must be such that they will furnish sufficient excitation to maintain normal voltage at the terminals of the generators when running at 50 per cent overload. Table XI gives the proper capacity of exciters for the generators listed. On account of the fact that the speed at which the unit runs is an important factor in the excitation required, no general figure can be given.

Exciters may be either direct-connected or belted to the shaft of the machine which they excite, or they may be separately driven. They are usually compound-wound and furnish current at 125 or 250 volts. Separately-driven exciters are preferred for most plants as they furnish a more flexible system, and any drop in the speed of the generator does not affect the exciter voltage. Ample reserve capacity of exciters should be installed, and in some cases storage batteries, used in conjunction with exciters, are recommended in order to insure reliability of service.

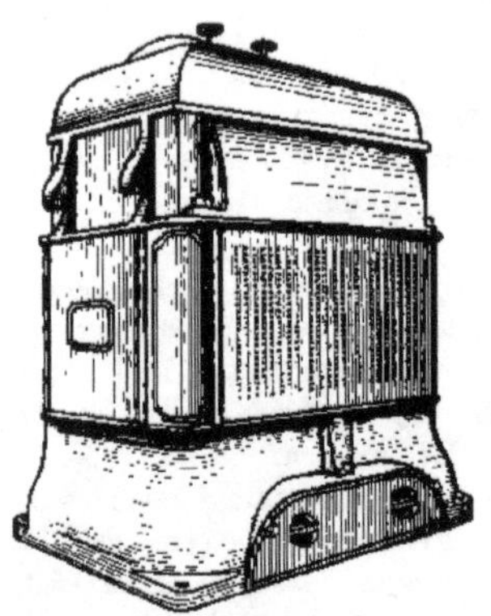

Fig. 14. Air-Cooled Transformer

Speed and Regulation. If direct-connected, the speeds of the generators will be determined by the prime mover selected. If belt-driven, small machines may be run at a high speed, as high-speed machines are cheaper than slow- or moderate-speed generators. In large sizes, this saving is not so great.

When shunt-wound dynamos are used, the inherent regulation should not exceed 2 to 3 per cent for large machines. For alternators, this is much greater and depends on the power factor of the load. A fair value for the regulation of alternators on non-inductive load is 10 per cent.

Mechanical Features. Motor-generator sets, boosters, frequency changers, and other rotating devices come under the head of special apparatus and are governed by the same general rules as generators.

Transformers. Transformers for stepping the voltage from that generated by the machine up to the desired line voltage, or *vice versâ*, at the substation, may be of three general types, according to the method of cooling. Large transformers require artificial means of cooling, if they are not to be too bulky and expensive. They may be air-cooled, oil-cooled, or water-cooled.

Fig. 15. Oil-Cooled Transformer

Air-cooled transformers, Fig. 14, are usually mounted over an air-tight pit fitted with one or more motor-driven blowers which feed into the pit. The transformer coils are subdivided so that no part of the winding is at a great distance from air and the iron is provided with ducts. Separate dampers control the amount of air which passes between the coils or through the iron. Such transformers give good satisfaction for voltages up to 20,000 or higher, and can be built for any capacity. Care must be taken to see that there is no liability of the air supply failing, as the capacity of the transformers is greatly reduced when not supplied with air.

Fig. 16. Water-Cooled Transformer

Oil-cooled transformers, Fig. 15, have their cores and windings placed in a large tank filled with oil. The oil serves to conduct the heat to the case, and the case is usually made either of corrugated sheet metal or of cast iron containing deep grooves, so as to increase the radiating surface. These transformers do not require such heavy insulation on the outside of the coils as air-blast machines because the oil serves this purpose. Simple oil-cooled transformers are seldom built for capacities exceeding 250 kw. as they become too bulky, but they are employed for the highest voltages now in use.

Fig. 17. 400-Kw. Water-Cooled Oil Transformer

Water-cooled transformers, Figs. 16 and 17, are used when high voltages are required. This type is like an oil-cooled transformer, but with water tubes arranged in coils in the top. Cold water passes through these tubes and aids in removing heat from the oil. Some types have the low-tension windings made up of tubes through which the water circulates. Water-cooled transformers must not have the supply of cooling water shut off for any length of time when under normal load or they will overheat.

One or more spare transformers should always be on hand and they should be arranged so that they can be put into service on very short notice.

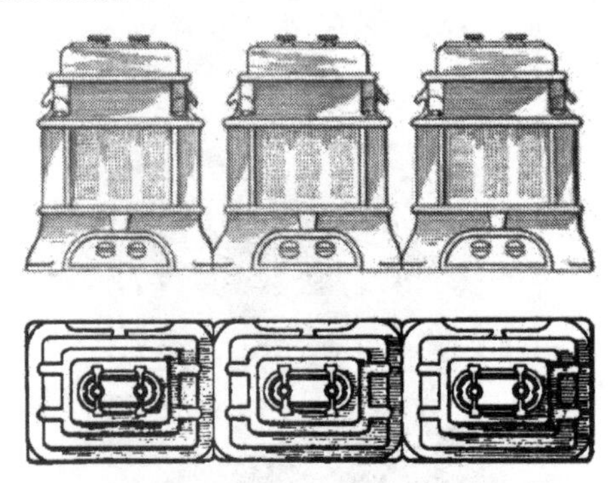

Fig. 18. Three-Phase Air-Blast Transformers. Total Capacity, 3,000 Kw.

Three-phase transformers allow a considerable saving in floor space, as shown by a comparison of the machines in Figs. 18 and

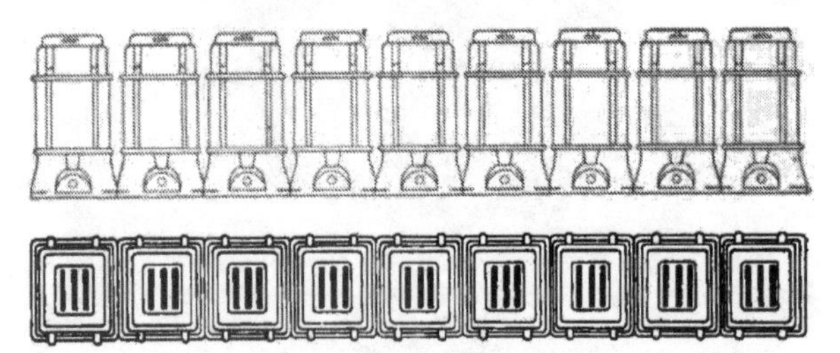

Fig. 19. Single-Phase Air-Blast Transformers. Total Capacity, 3,000 Kw.

19. They are cheaper than three separate transformers which make up the same capacity, but they are not as flexible as a single-phase transformer and one complete unit must be held for a reserve or "spare" transformer.

Storage Batteries. The use of storage batteries for central stations and substations is clearly outlined in "Storage Batteries."

The chief points of advantage are:

1. Reduction in fuel consumption due to the generating machinery being run at its greatest economy.
2. Better voltage regulation.
3. Increased reserve capacity and less liability to interruption of service.

The main disadvantage is the high first cost and depreciation.

SWITCHBOARDS

The switchboard is the most vital part of the whole system of supply, and should receive consideration as such. Its objects are: to collect the energy as supplied by the generators and to direct it to the desired feeders, either overhead or underground; to furnish a support for the various measuring instruments connected in service, as well as the safety devices for the protection of the generating apparatus; and to control the pressure of the supply. Some of the essential features of all switchboards are:

1. The apparatus and supports must be fire-proof.
2. The conducting parts must not overheat.
3. Parts must be easily accessible.
4. Live parts except for low potentials must not be placed on the front of the operating panels.
5. The arrangement of circuits must be symmetrical and as simple as it is convenient to make them.
6. Apparatus must be arranged so that it is impossible to make a wrong connection that would lead to serious results.
7. It should be arranged so that extensions may be readily made.

There are two general types—in the first, all of the switching and indicating apparatus is mounted directly on panels; and in the second, the current-carrying parts are at some distance from the panels, the switches being controlled by long connecting rods, or else operated electrically or by means of compressed air. The first may again be divided into direct-current and alternating-current switchboards. It is from the first class of apparatus that the switchboard gets its name and the term is still applied, even when the board proper forms the smallest part of the equipment. The term "switchgear" is now being introduced to cover all of the apparatus connected with the switching operations and the term "switchboard" is being reserved for the panels and their apparatus only. Switchboards have been standardized to the extent that standard generator, exciter, feeder, and motor panels may be purchased for certain classes

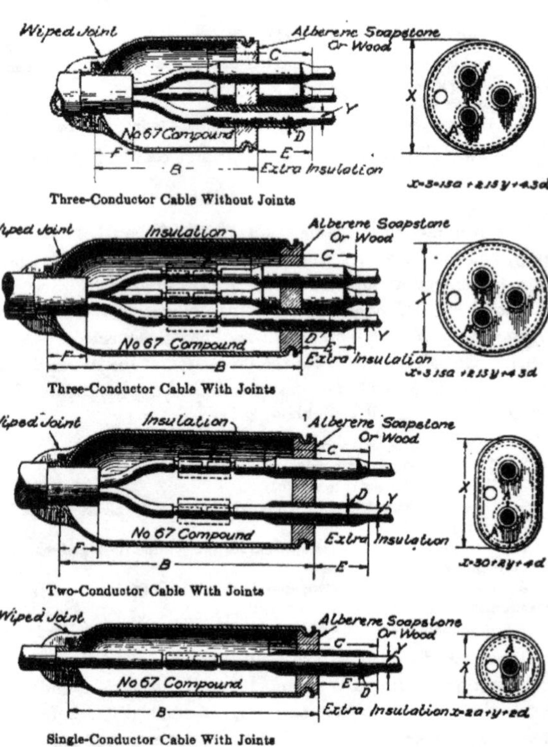

VOLTS	A	B	C	D	E	F
6 600	1	12	5	1/8	2½	1
13 200	1½	15	8	1/4	4	2
26 400	2	19	14	1/2	7	4

1/8-inch Lead or 1/16-inch Brass Bells

Fig. 20. Part Section—Showing Cable Bells in Place

THREE 6600-VOLT ALTERNATING-CURRENT GENERATORS INSTALLED FOR THE SANITARY DISTRICT OF CHICAGO

of work, but the vast majority of them are made up as semi-standard or special.

The leads which carry the current from the machines to the switches should be put in with very careful consideration. Their size should be such that they will not heat excessively when carrying the rated overload of the machine, and they should preferably be placed in fire-proof ducts, although low-potential leads do not always require this construction. Curves showing sizes for lead-covered cables for different currents are given in "Power Transmission." Table XII gives standard sizes of wires and cables together with the thickness of insulation necessary for different voltages. Cables should be kept separate as far as possible so that if a fault does occur on one cable, neighboring conductors will not be injured. For lamp and instrument wiring, such as leads to potential and current transformers, the following sizes of wire are recommended:

No. 16 or No. 14, wiring to lamp sockets.
No. 12 wire, $\frac{3}{64}''$ rubber insulation, all other small wiring under 600 volts potential.
No. 12, $\frac{3}{32}''$ rubber insulation for primaries of potential transformers from 600 to 3,500 volts.
No. 8, $\frac{5}{32}''$ rubber insulation for primaries of potential transformers up to 6,600 volts.
No. 8, $\frac{7}{32}''$ rubber insulation for primaries of potential transformers up to 10,000 volts.
No. 4, $\frac{21}{64}''$ rubber insulation for primaries of potential transformers up to 15,000 volts.
No. 4, $\frac{14}{32}''$ rubber insulation for primaries of potential transformers up to 20,000 volts.
No. 4, $\frac{17}{32}''$ rubber insulation for primaries of potential transformers up to 25,000 volts.

Where high-tension cables leave their metallic shields they are liable to puncture, so that the sheath should be flared out at this point and the insulation increased by the addition of compound. Fig. 20 shows such cable bells, as they are called, as are recommended by the General Electric Company. Other types of cable outlets are introduced from time to time. A very excellent type makes use of porcelain sleeve for each conductor at the point when it leaves the lead sheath.

Panels. Central-station switchboards are usually constructed of panels about 90 inches high, from 16 inches to 36 inches wide, and 1½ inches to 2 inches thick. Such panels are made of blue Vermont,

TABLE XII

Standard Wire
(Solid)

Area	Diameter Inches	Terminal Drilling	Amperes	Thickness of Rubber Insulation							Gauge
				Volts							
Circular Mils	Bare	Drill Number	Constant Current Capacity	600	3,500	6,600	10,000	16,000	20,000	25,000	B. & S.
2,582	.051	30	4								16
4,106	.064	30	6	3/64							14
6,530	.081	30	10	3/64	3/32						12
16,510	.128	18	25	3/64	3/32	5/32	7/32				8
26,251	.162	5	40	1/16							6
41,743	.204	1/4	60	1/16	3/32	5/32	7/32	21/64	11/32	13/32	4
66,373	.257	5/16	90	1/16							2
83,695	.289	11/32	110	5/64							1
105,593	.325	3/8	130	5/64	3/32	5/32	7/32	21/64	11/32	13/32	0
133,079	.365	7/16	170	5/64							00
167,805	.410	15/32	205	5/64							000
211,600	.460	17/32	250	5/64	3/32	5/32	7/32	21/64	11/32	13/32	0000

Standard Cable
(Stranded)

Circular Mils	Diameter, Inches Bare	Terminal Drilling Inches	Con. Cur. Capacity Amperes	Thickness of Rubber Insulation (For 6000 V. only)
250,000	.568	5/8	290	3/32
300,000	.637	21/32	240	3/32
350,000	.680	3/4	380	3/32
400,000	.735	13/16	420	3/32
500,000	.820	29/32	50)	3/32
600,000	.900	1	575	7/64
800,000	1.037	1 1/8	710	7/64
1,000,000	1.157	1 1/4	830	7/64
1,500,000	1.412	1 1/2	1100	1/8
2,000,000	1.65	1 3/4	1350	1/8

pink Tennessee, or white Italian marble, or of black enameled or oiled slate. Slate is not recommended for voltages exceeding 1,100. The panels are made in two or three parts. When made in two parts, the sub-base is from 24 to 28 inches high. They are polished on the front and the edges are beveled. Angle and tee bars or pipe work, together with foot irons and tie rods, form the supports for such panels, and on these panels are mounted the instruments, main switches, or controlling apparatus for the main switches, as the case may be, together with relays and hand wheels for rheostats and regulators.

The usual arrangement of the panels is to have a separate panel for each generator, exciter, and feeder, together with what is known as a station or total-output panel. In order to facilitate extensions and simplify connections, the feeder panels are located at one end of the board, the generator panels are placed at the other end, and the total-output panel occupies a position between the two. The main bus bars extend throughout the length of the generator and feeder panels, and the desired connections are readily made. The instruments required are very numerous. Lists of meters required for standard practice and regular panels are given later.

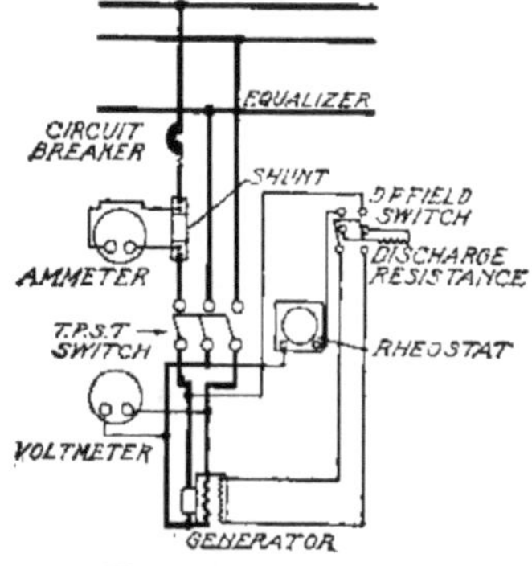

Fig. 21. Wiring Diagram of D. C. Generator Panel

For *direct-current generator panels* or for the *direct-current side of synchronous converters, two-wire system,* there are usually required:

1 Main switch
1 Field switch
1 Ammeter
1 Voltmeter
1 Field rheostat with controlling mechanism
1 Circuit breaker
1 4-point starting switch (for use when machine is to be started as a direct-current motor).
Bus bars and various connections.

These may be arranged in any suitable order, the circuit breaker being preferably located at the top so that any arcing which may occur will not injure other instruments. Fig. 21 gives a wiring diagram of such a panel.

The main switch may be a single- or a double-throw, depending on whether one or two sets of bus bars are used. It may be a triple-pole, as shown in Fig. 21, in which the middle bar serves as the equalizing switch, or the equalizing switch may be mounted on a pedestal near the machine, in which case the generator switch would be double-pole.

The field switch for large machines should be double-pole fitted with carbon breaks and arranged with a discharge resistance consisting of a resistance which is thrown across the terminals of the

Fig. 22. Carbon Break Circuit Breaker

field just before the main circuit is opened. One voltmeter located on a swinging bracket at the end of the panel, and arranged so that it can be thrown across any machine or across the bus bars by means of a dial switch, is sometimes used, but it is preferable to have a separate meter for each generator.

Small rheostats are mounted on the back of the panel, but large ones are chain-operated and preferably located below the floor, the controlling hand wheel being mounted on the panel.

The circuit breaker may be of the carbon break or the magnetic blow-out type. Figs. 22 and 23 show circuit breakers of these types. Lighting panels for low potentials are often fitted with fuses instead of circuit breakers, in which case they may be open fuses on the back

Fig. 23. Magnetic Blow-Out Circuit Breaker

of the panel or enclosed fuses on either the front or back of the panel.

A panel for a *direct-current generator or a synchronous converter for a 3-wire system* should contain:

2 Ammeters.
2 Circuit breakers. Fuses used on small generators.

3 Single-pole switches, double-throw if there are two sets of bus bars. For a three-wire generator or a synchronous converter two single-pole or one double-pole switch may be used, in which case the neutral wire is not brought to the switchboard.
2 Hand-wheels for the field rheostats. But one required if a three-wire generator is used but the two are necessary if the three-wire system is obtained by the use of two generators or a balancer set.
2 Field switches. But one is required for a three-wire generator or synchronous converter.
1 Four-point starting switch. Required only when the machine is to be started as a direct-current motor at times.
2 Potential receptacles, four-point, used in connection with a voltmeter, usually mounted on a swinging bracket. Only one is required for the three-wire generator or the synchronous converter.

An alternating-current generator or a synchronous motor panel for a *three-phase, three-wire system* will require:

3 Ammeters. Only one required for a single-phase panel or for a synchronous motor.
1 Three-phase indicating wattmeter.
1 Voltmeter.
1 8-point potential receptacle used to connect the above voltmeter across any phase. Not necessary for the synchronous motor.
1 Field ammeter. Convenient but not always necessary.
1 Double-pole field switch with discharge clips.
1 Hand-wheel for field rheostat.
1 Synchronizing receptacle.
1 Triple-pole oil switch, usually non-automatic for generators but automatic for motors. This may be single- or double-throw, depending upon the bus bar arrangements.
1 Synchronizer. A single instrument may serve for several machines.
2 Current transformers.
2 Potential transformers. Only one necessary for motor.
1 Power Factor indicator. Not always necessary.
1 Governor control switch. Not always necessary.

Where the switches are of the remote control type, the control switches or the operating handles are mounted on the panel.

A three-phase induction-motor panel should contain:

1 Ammeter.
1 Automatic oil switch, preferably operated by means of an inverse time-limit relay.
The starting compensator used with induction motors is usually mounted independently of the switchboard panel.

The instruments used on a *synchronous converter panel, alternating-current control,* are:

1 Ammeter.
1 Synchronizing receptacle.
1 Oil-switch, automatic.
1 Potential transformer.
2 Current transformers.
1 Switch for control of regulator where a regulator is used and operated by means of a small motor.

A three-phase feeder panel requires:

3 Ammeters. In some cases only one is necessary.
1 Automatic oil switch.
2 Current transformers.
1 Potential Transformer. Not always needed.
1 Voltmeter. Not always needed.
1 Hand-wheel for control of regulator where a regulator is used.

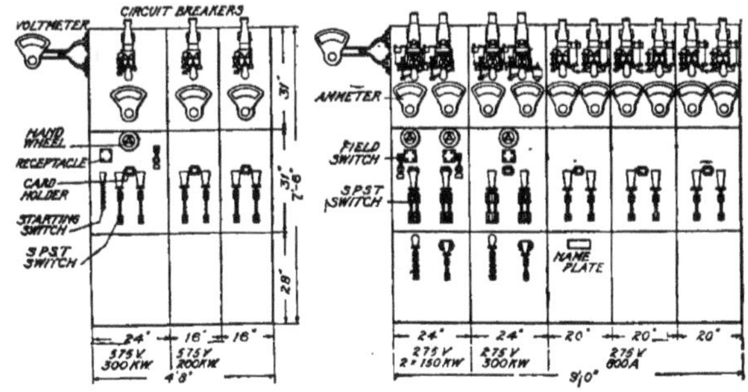

Fig. 24. Standard Switchboard Panel

Direct-Current feeder panels contain:

1 Ammeter. Two are required for a 3-wire feeder.
1 Circuit breaker, single-pole. Two are required for a 3-wire feeder.
1 or more main switches, single-pole or double-pole, and single- or double-throw, depending upon the number of bus bars.
1 Recording wattmeter, not always used.
1 Potential receptacle.
Apparatus for controlling regulators when such are used.

One voltmeter usually serves for several feeder panels, such a meter being mounted above the panels or on a swinging bracket at the end. Switches should preferably be of the quick-break type. Figs. 24 and 25 show some standard switchboard panels as manufactured by the General Electric Company.

Exciter panels are nothing more than generator panels on a small scale. The necessary instruments for a panel controlling one exciter are:

1 Ammeter.
1 Field rheostat.
1 Voltmeter.
1 3-pole switch with fuses.
1 4-point potential receptacle.
1 Equalizing rheostat. This is necessary only where a Tirrill regulator is used and more than one exciter is operated on the same set of excitation buses.

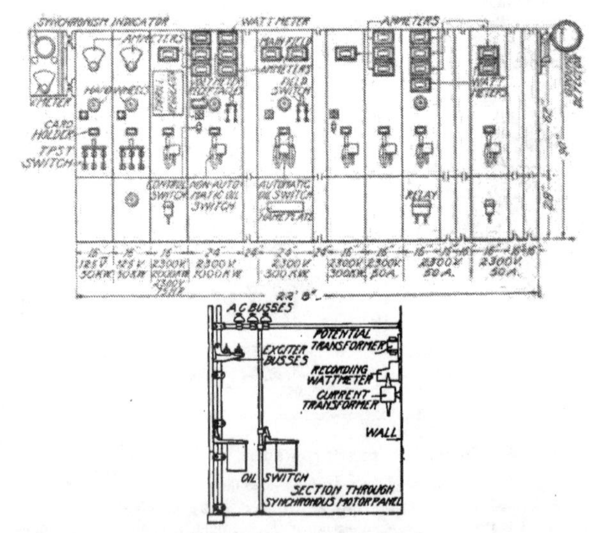

Fig. 25. Standard Switchboard Panel

Total Output Panels contain instruments recording the total power delivered by the plant to the switchboard. The paralleling of alternators is treated in "Management of Dynamo Electric Machinery."

For the higher voltages on alternating-current boards, the measuring instruments are no longer connected directly in the circuit, and the main switch is not mounted directly on the panel. Current and potential transformers, as called for in the lists given in connection with the different panels, are used for connecting to the indicating voltmeters and the ammeters and the recording wattmeters, and potential transformers are used for the synchronizing device.

Fig. 26. Three-Phase Oil Switch with Oil Container Removed

These transformers are mounted at some distance from the panel, while the switches may be located near the panel and operated by a system of levers, or they may be located at a considerable distance and operated by electricity or by compressed air.

Oil Switches. Oil switches are recommended for all high potential work for the following reasons:

By their use it is possible to open circuits of higher potential and carrying greater currents than with any other type of switch.

They may be made quite compact.

They may readily be made automatic and thus serve as circuit breakers for the protection of machines and circuits when overloaded.

There are several types of oil switches on the market. A switch constructed for three-phase work, to be closed by hand and to be

Fig. 27. Three-Phase Oil Switch with Oil Container in Place

electrically tripped or opened by hand, is shown in Fig. 26. This shows the switch without the can containing the oil. Fig. 27 shows a similar switch hand-operated, with the can in place. Both of these switches are arranged to be mounted on the panel. Fig. 28 shows how the same switches are mounted when placed at some distance from the panel. For high voltages, they are placed in brick

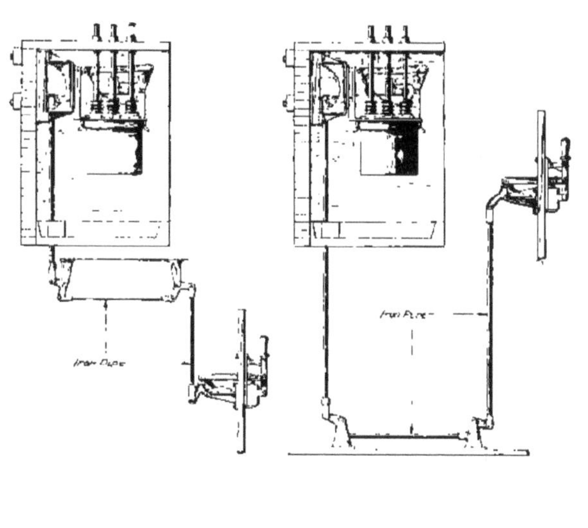

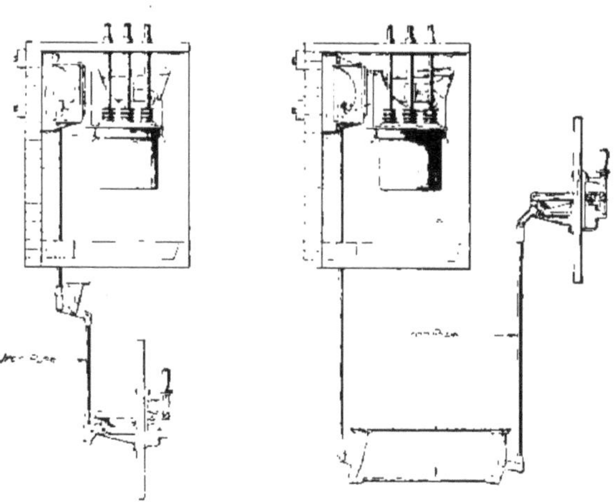

Fig. 28. Four Arrangements of Oil Switch when Mounted at Some Distance from the Panel

cells and often three separate single-pole switches are used, each placed in a separate cell so that injury to the contacts in one leg will in no way affect the other parts of the switch. A form of oil switch used for the higher potentials and currents met with in practice, is shown in Fig. 29. This particular switch is operated by means of an electric motor, though it may as readily be arranged to operate by means of a solenoid or by compressed air. General

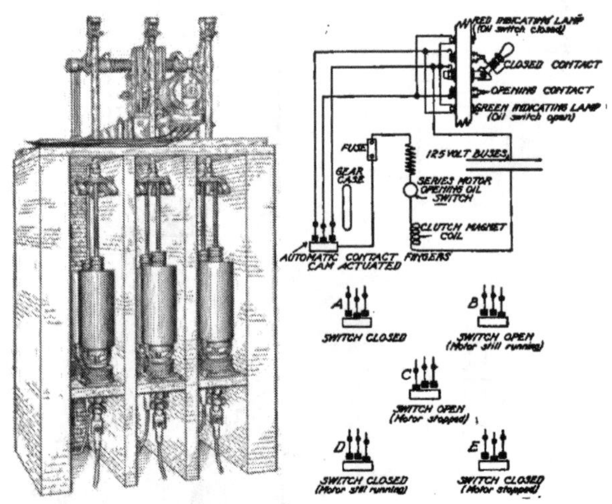

Fig. 29. Form of Oil Switch for High Potentials

practice is to place all high-tension bus bars and circuits in separate compartments formed by brick or cement, and duplicate bus bars are quite common.

Oil switches are made automatic by means of *tripping magnets*, which are connected in the secondary circuits of current transformers, or they may be operated by means of relays fed from the secondaries of current transformers in the main leads. Such relays are made very compact and can be mounted on the front or back of the switchboard panels. The wiring of such tripping devices is shown in Fig. 30.

With remote control of switches, the switchboard becomes in many instances more properly a switch house, a separate building being devoted to the bus bars, switches and connections. In other cases a framework of angle bars or gas pipe is made for the support of the switches, bus bars, current and potential transformers, etc. The supports for the controlling switches are sometimes mounted in a nearly horizontal position, forming the bench type of control board.

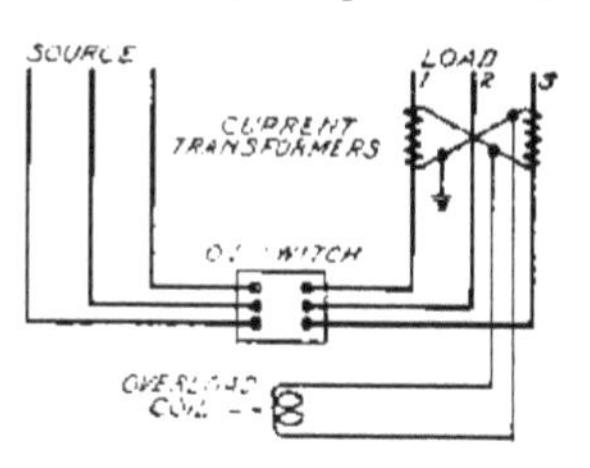

Additional types of panels which may be mentioned are transformer panels, usually containing switching apparatus only, and arc-board panels. The latter are arranged to operate with plug switches. A single panel used in the operation of series transformers on arc-lighting circuits is shown in Fig. 31.

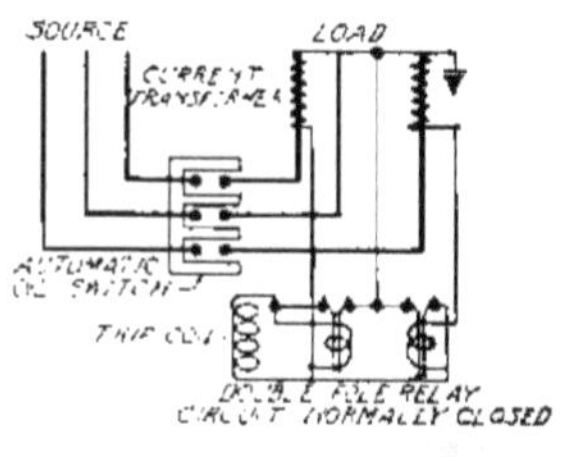

Safety Devices. In addition to the ordinary overload tripping devices which have already been considered, there are various safety devices necessary in connection with the operation of central stations. One of the most important of these is the *lightning arrester*. For direct-current work, the lightning arrester often takes the form of a single gap connected in series with a high resistance and fitted with some device for destroying the arc formed by discharge to the ground. One of these is connected between either side of the circuit and the ground, as shown diagrammatically in

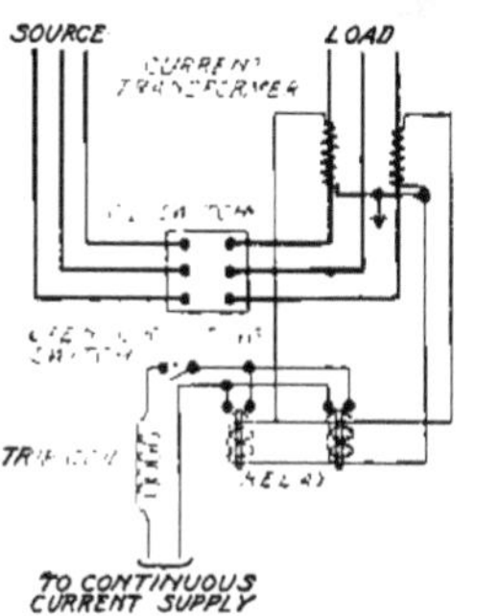

Fig. 30. Wiring Diagram for Tripping Devices

Fig. 32. A "kicking" coil is connected in circuit between the arresters and the machine to be protected, to aid in forcing the lightning discharge across the gap. In railway feeder panels such kicking coils are mounted on the backs of the panels.

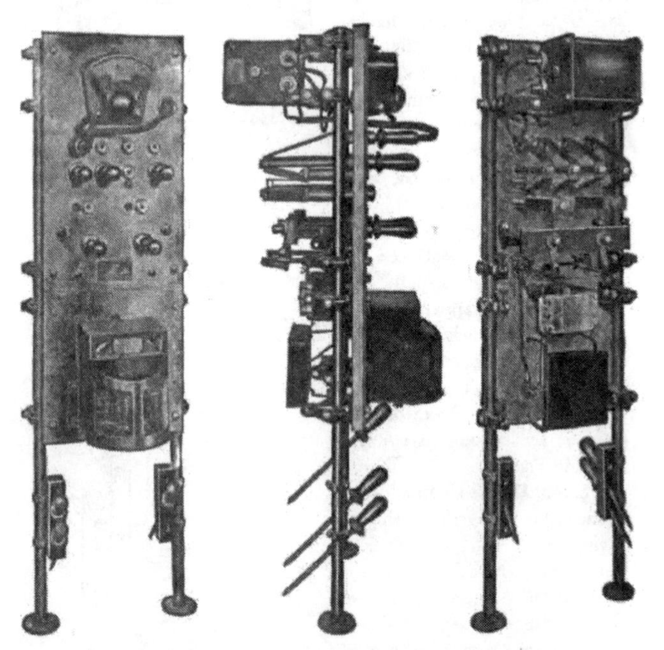

Fig. 31. Single Panel for Series Transformers in Arc-Lighting Circuits

For alternating-current work, several gaps may be arranged in series, these gaps being formed between cylinders of "non-arcing" metal. High resistances and reactance coils are used with these, as in direct-current arresters. Fig. 33 shows connections for a 10,000-volt lightning arrester. The resistance used in connection with lightning arresters are of special design and non-inductive. In recent types these resistances are connected in shunt to the gaps as shown in Fig. 34. Lightning arresters should always be provided with knife blade switches so that they can be disconnected from the circuit for inspection and repairs. A typical installation of lightning arresters is shown in Fig. 35.

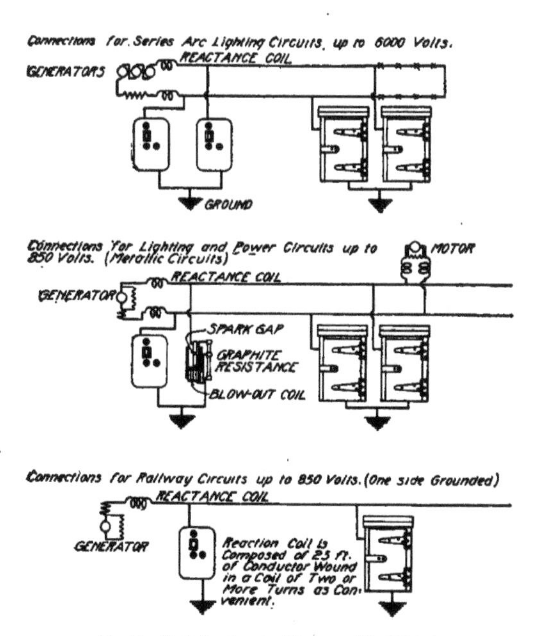

Fig. 32. Lightning-Arrester Diagrams of D. C. Work

In place of a series of gaps a single gap with terminals made in the form of horns is employed in some cases for lightning protection. Such an arrester is known as a *horn gap*, or *horn arrester*. The gap is connected between the line and the ground and when the potential strain becomes great enough the gap is broken down. The arc formed by the machine current after the gap is broken down

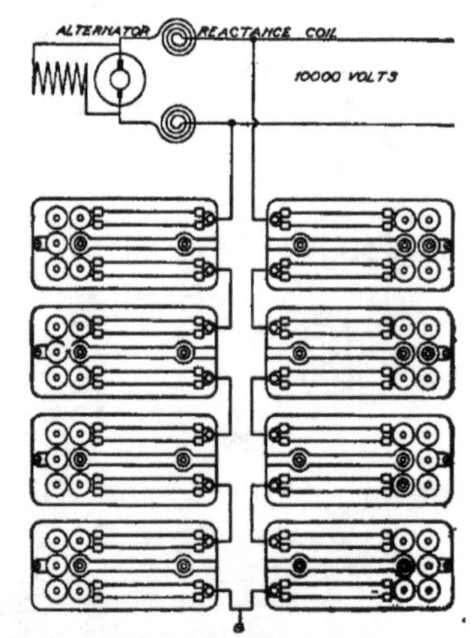

Fig. 33. Connections for 10,000-Volt Lightning Arrester

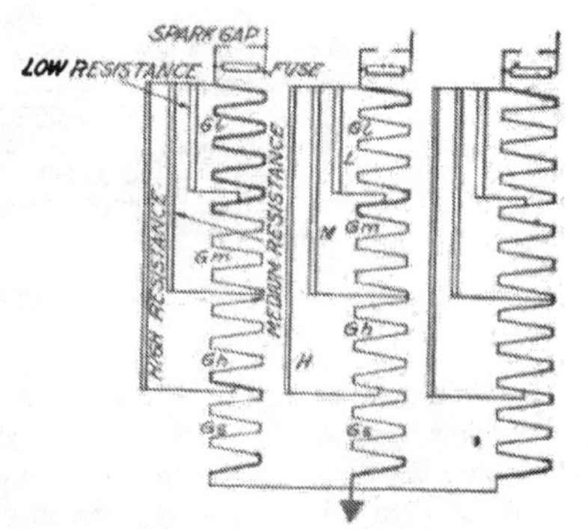

Fig. 34. Non-Inductive Resistances in Shunt with Lightning Arrester

Fig. 35. Typical Installation of Lightning Arresters

rises and lengthens until it can no longer be maintained by the generator or generators in service. The horn arrester as applied to series lighting circuits is shown in Fig. 36. A series resistance, shown in the lower part of the figure, is used with this particular arrester.

Fig. 36. "Horn" Lightning Arrester

The most recent development in the way of lightning protection is the introduction of the aluminum cell arrester. The elementary cell consists of two aluminum plates, on which a film of aluminum hydroxide is formed, and which are immersed in a suitable electrolyte. The peculiar property of such a cell which makes it useful as a lightning arrester is that it has a high resistance up to a certain potential impressed upon it but when a critical value of voltage is reached, the resistance becomes very low. The critical voltage for a single unit for alternating current is between 335 and 360 volts, and such a cell may be connected to a 300-volt circuit with only a very small current flow. For higher voltages, a number of cells are connected in series and a horn gap is inserted between the arrester and the line wire. The gap prevents any flow of current unless the arrester is brought into action by the discharge of excess line potential, in which case the aluminum cells offer a path of low resistance for the discharge of potential so long as the voltage is in excess of the critical voltage, but the machine or line potential, which is below the critical voltage of the arrester, cannot force enough current through the arrester circuit to maintain the arc at the gap. There is some dissolution of the film of hydroxide if the cell is left standing and not connected to the circuit, but it is readily formed again when the circuit is made. Arresters using a gap should have the gap closed for a short interval daily in order to insure a proper film on the aluminum plates. Views of the aluminum arrester are shown in Figs. 37 and 38.

Reverse-current relays are installed when machines or lines

are operated in parallel. If two or more alternators are running and connected to the same set of bus bars, and one of these should fail to generate voltage by the opening of the field circuit, or some other cause, the other machine would feed into this generator and might cause considerable damage before the current flowing would be sufficient to operate the circuit breaker by means of the overload trip coils. To avoid this, reverse-current relays are used. They are so

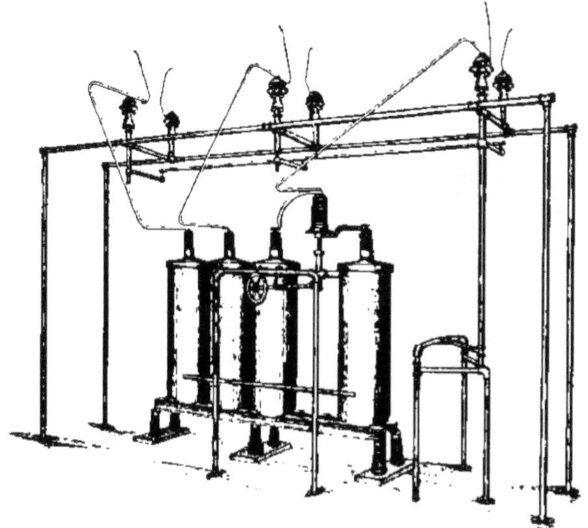

Fig. 37. Installation of Aluminum Lightning Arrester for 35,000 Volts

arranged as to operate at, say $\frac{1}{4}$ the normal current of the machine or line, but to operate only when the power is being delivered in the wrong direction.

Speed-limit devices are used on both engines and rotary converters to prevent racing in the one case and running away in the second. Such devices act on the steam supply of engines and on the direct-current circuit breakers of rotary converters, respectively.

Complete wiring diagrams for standard switchboards are shown in Figs. 39 and 40.

SUBSTATIONS

Substations are for the purpose of transforming the high potentials down to such potentials as can be used on motors or lamps, and in many cases to convert alternating current into direct current.

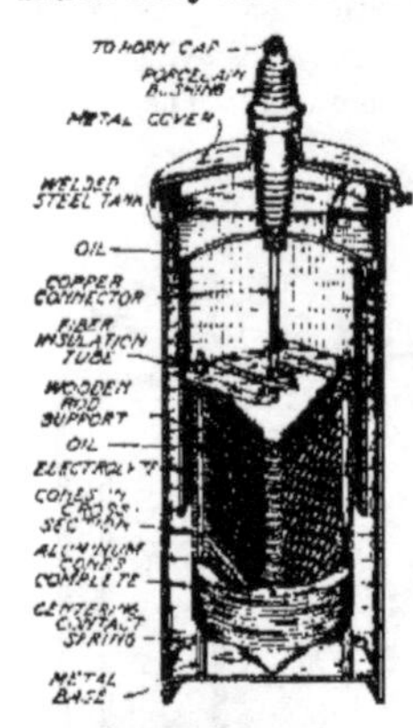

Fig. 38. Cross-Section of Aluminum Arrester

Step-down transformers do not differ in any respect from step-up transformers. Either motor-generator sets or rotary converters may be used to change from alternating to direct current. The former consist of synchronous or induction motors, direct connected to direct-current generators, mounted on the same bedplate. The generator may be shunt or compound wound, as desired. Rotary or synchronous converters are direct-current generators, though specially designed; and they are fitted with collector rings attached to the winding at definite points. The alternating current is fed into these rings and the machine runs as a synchronous motor, while direct current is delivered at the commutator end. There is a fixed relation between the voltage applied to the alternating-current side and the direct-current voltage, which depends on the shape of the wave form, losses in the armature, pole pitch of the machine, method of connection, etc. The generally accepted values are given in Table XIII.

The increase of capacity of six-phase machines over other machines of the same size is given in Table XIV.

This increase is due to the fact that, with a greater number of phases, less of the winding is traversed by the current which passes through the converter. The saving by increasing the number of phases beyond six is but slight and the system becomes too complex. Rotary converters may be over-compounded by the addition of series fields, provided the reactance in the alternating circuits be of a proper value. It is customary to insert reactance coils in the leads from the low-tension side of the step-down transformers to the collector rings to bring the total reactance to a value which will insure

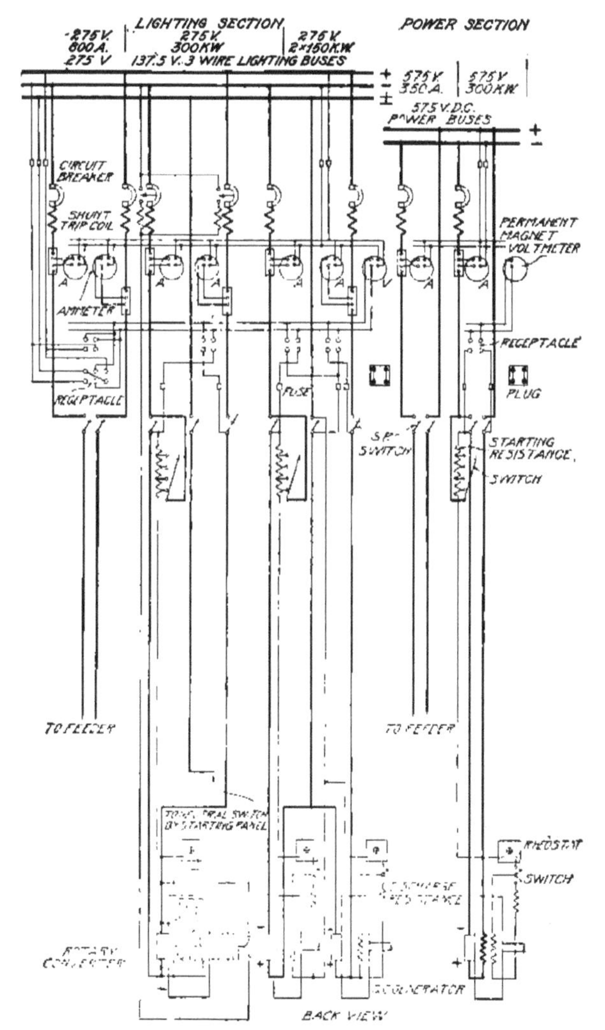

Fig. 39. Wiring Diagram for Standard Switchboard

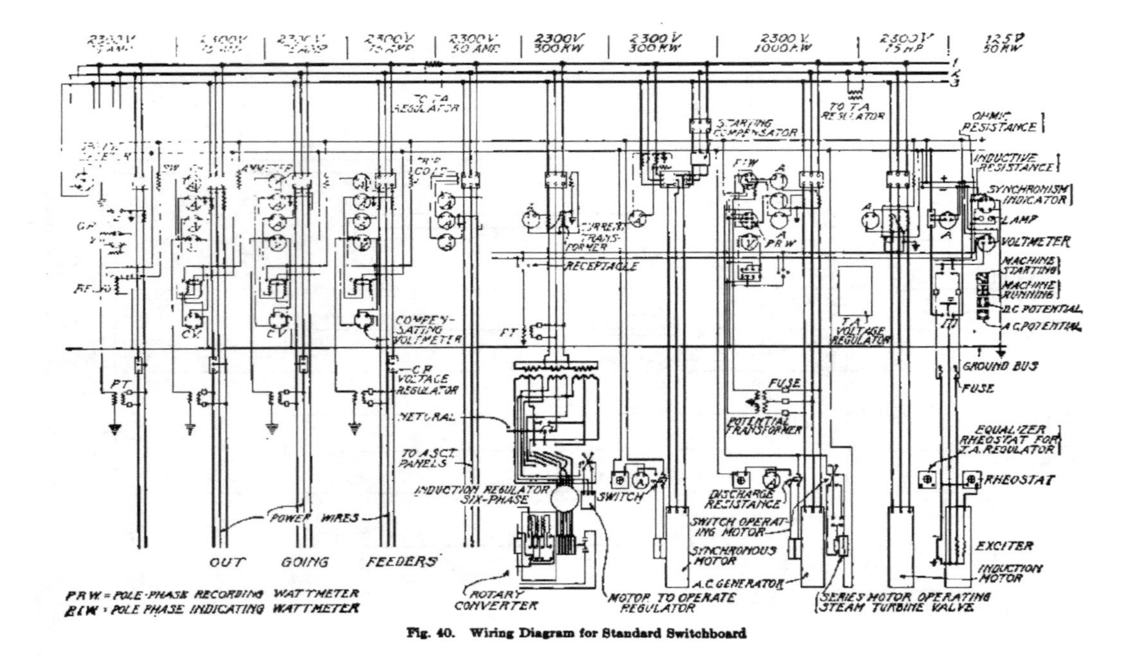

Fig. 40. Wiring Diagram for Standard Switchboard

TABLE XIII

Full Load Ratios

Current		Potential Per Cent
Continuous		100
Two-phase	550 volts	72.5
and Six-phase	250 volts	73
(diametrical)	125 volts	73.5
Three-phase	550 volts	62
and Six-phase	250 volts	62
(Y or delta)	125 volts	63

the desired compounding. Again, the voltage may be controlled by means of induction regulators placed in the alternating-current leads.

Two other methods of potential regulation for synchronous converters are in use. In the first of these methods a series generator is used, this generator consisting of a polyphase armature attached to the rotary converter shaft and revolving in a separate field. The phases of this armature are connected between the collector rings of the machine and the taps to the converter armature, and the voltage impressed upon the converter taps amounts to the line voltage plus or minus the potential developed in the regulating armature. By means of a suitable field rheostat for the regulating field, any ordinary range of direct current at the brushes of the converter can be obtained with a constant voltage of alternating-current supply. Fig. 41 shows a converter equipped with this regulating device.

In the regulating-pole converter each pole of the machine is made up of two parts, the main pole and the regulating pole. By

TABLE XIV

Capacity Ratios

Continuous-current generator	100
Single-phase converter	85
Two-phase converter	164
Three-phase converter	134
Six-phase converter	196

varying the excitation of the regulating pole the ratio of conversion between the alternating-current voltage and the direct-current voltage can be changed through a considerable range—a sufficient range to cover the requirements ordinarily required in practice. Fig. 42 shows a view of a regulating pole converter. Motor-generators are more costly and occupy more space than rotary converters but the regulation of the voltage is much better and they are to be preferred for lighting purposes.

BUILDINGS

The power station usually has a building devoted entirely to this work, while the substations, if small, are often made a part of

Fig. 41. Rotary Converter Equipped with Regulating Generator

other buildings. While the detail of design and construction of the buildings for power plants belongs primarily to the architect, it is the duty of the electrical engineer to arrange the machinery to the best advantage, and he should always be consulted in regard to the general plans, at least, as this may save much time and expense in the way of necessary modifications. The general arrangement of the machinery will be taken up later, but a few points in connection with the construction of the buildings and foundations will be considered here.

Space must be provided for the boilers—this may be a separate building—engine and dynamo room, general and private offices, store rooms and repair shops. Very careful consideration should be given to each of these departments. The boiler room should be parallel with the engine room, so as to reduce the necessary amount of steam piping to a minimum, and if both rooms are in the same building a brick wall should separate the two, no openings which would allow dirt to come from the boiler room to the engine room being allowed. The height of both boiler and engine rooms should

Fig. 42. Regulating-Pole Rotary Converter

be such as to allow ample headway for lifting machinery and space for placing and repairing boilers, while provision should be made for extending these rooms in at least one direction. Both engine and boiler rooms should be fitted with proper traveling cranes to facilitate the handling of the units. In some cases the engines and dynamos occupy separate rooms, but this is not general practice. Ample light is necessary, especially in the engine rooms. The size of the offices, store rooms, etc., will depend entirely on local conditions.

TABLE XV

Thickness of Walls for Power Plants

Width of Building Clear Between Walls	Height of Wall	First Section		Second Section		Third Section	
		Height	Thickness	Height	Thickness	Height	Thickness
25 feet	40 feet	40 feet	12 inches				
25 feet	40-60 feet	40 feet	16 inches	To top	12 inches		
25 feet	60-75 feet	25 feet	20 inches	To top	16 inches		
25 feet	75-85 feet	20 feet	24 inches	20-60 ft.	20 inches	To top	16 inches
25 feet	85-100 ft.	25 feet	28 inches	25-50 ft.	24 inches	50-75 feet	20 inches

NOTE. With clear space exceeding 25 feet the walls should be made 4 inches thicker for each 10 feet or fraction thereof in excess of 25 feet. For buildings greater than 100 feet in height, each additional 25 feet or fraction thereof next above the curb shall be increased 4 inches in thickness.

Foundations. The foundations for both the walls and the machinery must be of the very best. It is well to excavate the entire space under the engine room to a depth of eight to ten feet so as to form a basement, while in most cases the excavations must be made to a greater depth for the walls. Foundation trenches are sometimes filled with concrete to a depth sufficient to form a good underfooting. The area of the foundation footing should be great enough to keep the pressure within a safe limit for the quality of the soil. The walls themselves may be of wood, brick, stone, or concrete. Wood is used for very small stations only, while brick may be used alone or in conjunction with steel framing, the latter construction being used to a considerable extent. If brick alone is used, the walls should never be less than twelve inches thick, and eighteen to twenty inches is better for large buildings. They must be amply reinforced with pilasters. Stone is used only for the most expensive stations.

Table XV, which is taken from the New York Building Laws, may serve as a guide to the thickness of walls for power plants. The interior of the walls is formed of glazed brick, when the expense of such construction is warranted. In fireproof construction, which is always desirable for power stations, the roofs are supported by steel trusses and take a great variety of forms. Fig. 43 shows what has been recommended as standard construction for lighting stations, showing both brick and wood construction. The floors of the engine room should be made of some material which will not form grit or dust. Hard tile, unglazed, set in cement or wood floors, is desirable. Storage battery rooms should be separate from all others and should have their interior lined with some material which will not be affected by the acid fumes. The best of ventilation is desirable for all parts of the station, but is of particular importance in the dynamo room if the machines are being heavily loaded. Substation construction does not differ from that of central stations when a separate building is erected. They should be fireproof if possible.

The foundations for machinery should be entirely separate from those of the building. Not only must the foundations be

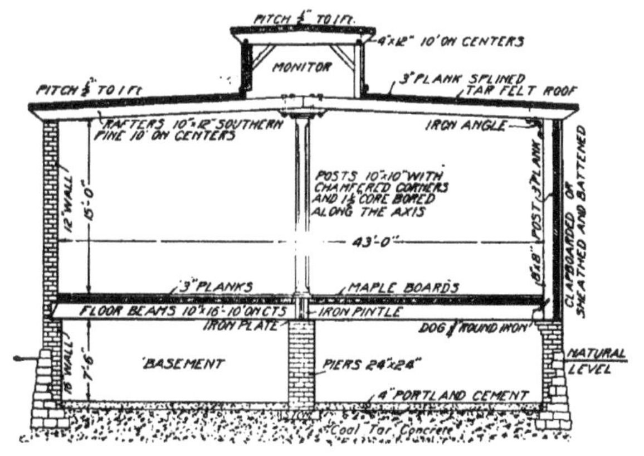

Fig. 43. Standard Construction for Lighting Station. Brick and Wood Construction

stable, but in some locations it is particularly desirable that no vibrations be transmitted to adjoining rooms and buildings. A loose or sandy soil does not transmit such vibrations readily, but firm earth or rock transmits them almost perfectly. Sand, wool, hair, felt, mineral wool, and asphaltum concrete are some of the materials used to prevent this. The excavation for the foundation is made from 2 to 3 feet deeper and 2 to 3 feet wider on all sides than the foundation, and the sand, or whatever material is used, occupies this extra space.

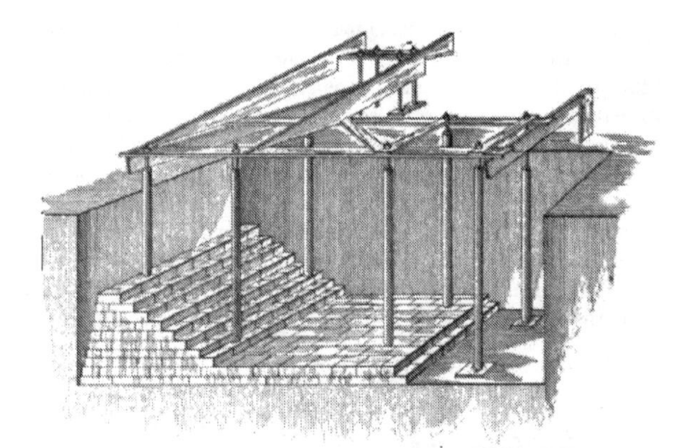

Fig. 44. Foundation for Machines Showing Use of Template and Iron Pipe for Holding Bolts

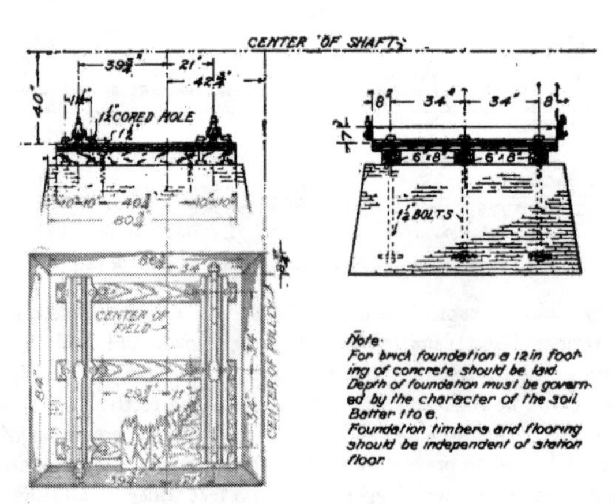

Fig. 45. Foundation for 150-Kw. Generator

Brick, stone, or concrete is used for building up the greater part of machinery foundations, the machines being held in place by means of bolts fastened in masonry. A template, giving the

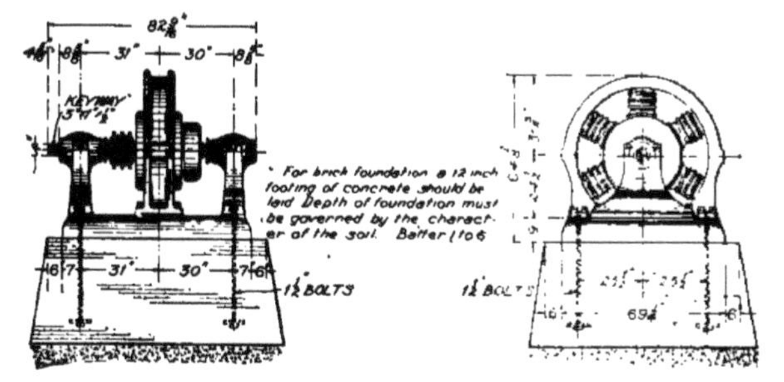

Fig. 46. Foundation for Rotary Converter

location of all bolts to be used in holding the machine in place, should be furnished, and the bolts may be run inside of iron pipes with an internal diameter a little greater than the diameter of the

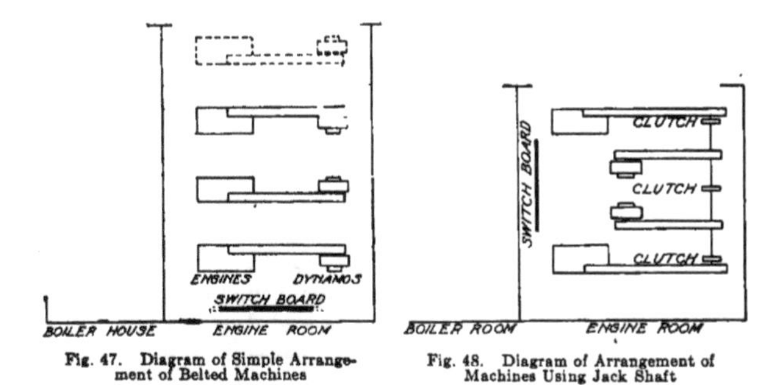

Fig. 47. Diagram of Simple Arrangement of Belted Machines

Fig. 48. Diagram of Arrangement of Machines Using Jack Shaft

bolt. This allows some play to the bolt and is convenient for the final alignment of the machine. Fig. 44 gives an idea of this construction. The brickwork should consist of hard-burned brick of the best quality, and should be laid in cement mortar. It is well to fit brick or concrete foundations with a stone cap, forming a level surface on which to set the machinery, though this is not necessary. Generators are sometimes mounted on wood bases to furnish insulation for the frame. Fig. 45 shows the foundation for a 150-kw. generator, while Fig. 46 shows the foundation for a rotary converter.

Station Arrangement. A few points have already been noted in regard to station arrangement, but the importance of the subject demands a little further consideration. Station arrangement depends chiefly upon two facts—the location and the machinery to be installed. Undoubtedly the best arrangement is with all of the machinery on one floor with, perhaps, the operating switchboard mounted on a gallery so that the attendants may have a clear view of all the machines. Fig. 47 shows the simplest arrangement of a plant using belted machines. Fig. 48 shows an arrangement of units where a jack shaft is used. Direct-current machines should be placed so that the brushes and commutators are easily accessible and the switchboard should be placed so as not to be liable to accidents, such as the breaking of a belt or a flywheel.

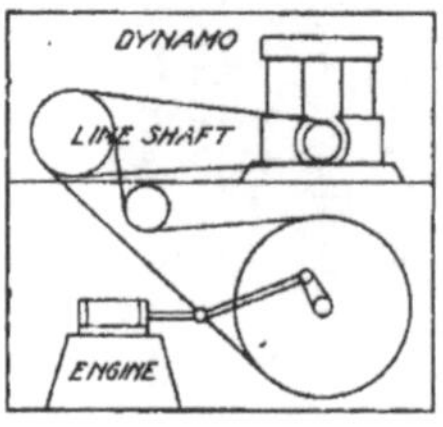

Fig. 49. Diagram of Double-Deck Arrangement of Machines

When the cost of real estate prohibits the placing of all of the machinery on one floor, the engines may be placed on the first floor and belted to generators on the second floor, Fig. 49. It is always desirable to have the engines on the main floor, as they cause considerable vibration when not mounted on the best of foundations. The boilers, while heavy, do not cause such vibration and they may be placed on the second or third floor. Belts should not be run vertically, as they must be stretched too tightly to prevent slipping.

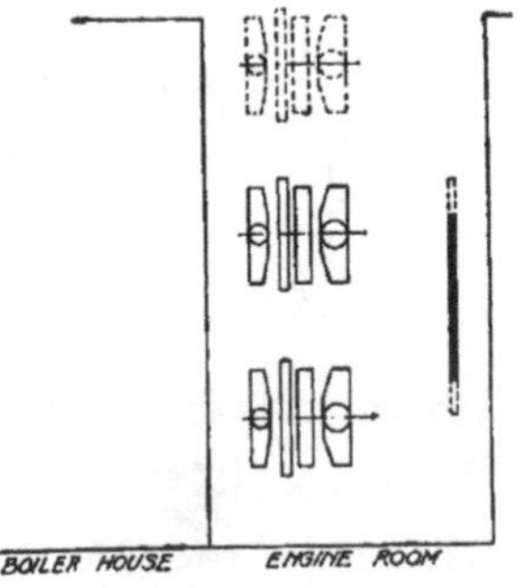

Fig. 50. Diagram of Station Using Direct-Connected Units

Fig. 50 shows a large station using direct-connected units, while Figs. 51 and 52 show the arrangement of the turbine plant of the Boston Edison Electric Illuminating Company. This station will contain twelve large turbine units when completed. Note the arrangement of boilers when several units are required for a single prime mover.

The use of the steam turbine has led to the introduction of a type of station known as a *double-deck power plant* and used in some instances where it is desirable to save space. In this type of station the boilers are placed on the ground floor and the turbines, which are of the horizontal type, are placed on a second floor directly above the boilers. Since there is but little vibration to the turbines and only light foundations are necessary, this construction may be readily carried out. Fig. 53 shows the general arrangement of such a plant. The use of a separate room or building for the cables, switches, and operating boards is becoming quite common for high-tension generating plants. The remarkable saving in floor space brought about by the turbine is readily seen from Fig. 54. The total floor space occupied by the 5,000 kw. units of the Boston station is 2.64 square feet per kw. This includes boilers—of which there are eight, each 512 h. p. for each unit—turbines, generators, switches, and all auxiliary apparatus. For the 10,000 and 15,000 kw. turbine sets now coming into use, this figure is still further reduced.

When transformers are used for raising the voltage, they may be placed in a separate building, as is the case at Niagara Falls, or the transformers may be located in some part of the dynamo room, preferably in a line parallel to the generators.

Fig. 55 shows the arrangement of units in a hydraulic plant. Fig. 56 is a good example of the practice in substation arrangement. The switchboard is at one end of the room, while the rotary converters and transformers are along either side.

Large cable vaults are installed at the stations operating on underground systems, the separate ducts being spread out, and sheet-iron partitions erected to prevent damage being done to cables which were not originally defective, by a short circuit [in any one feeder.

Station Records. In order to accurately determine the cost of generating power and to check up on uneconomical or improper methods of operation and lead to their improvement, accurately-kept station records are of the utmost importance. Such records

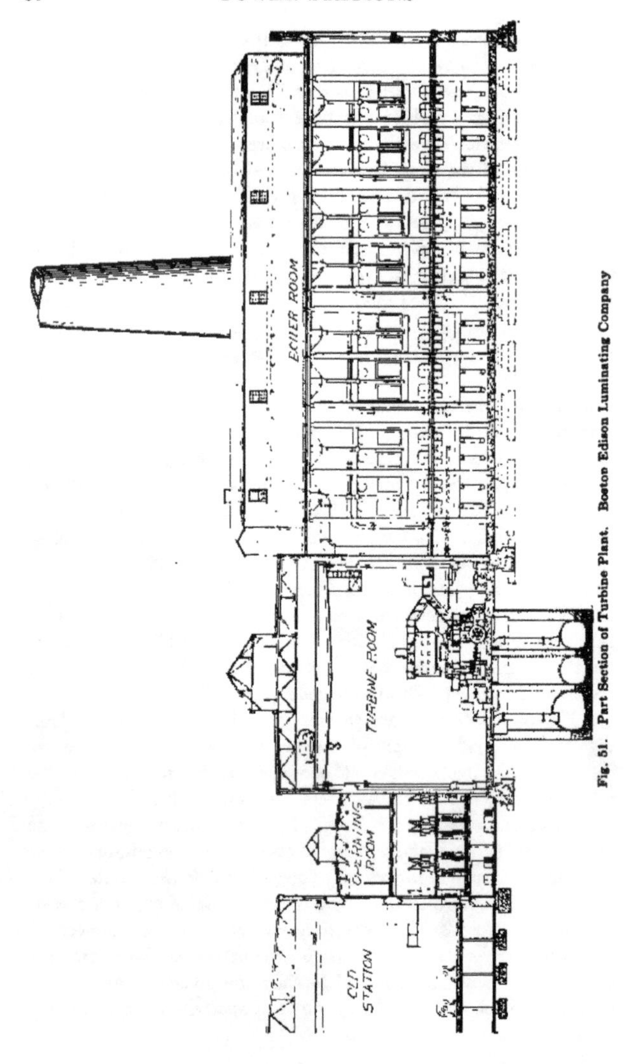

Fig. 51. Part Section of Turbine Plant. Boston Edison Luminating Company

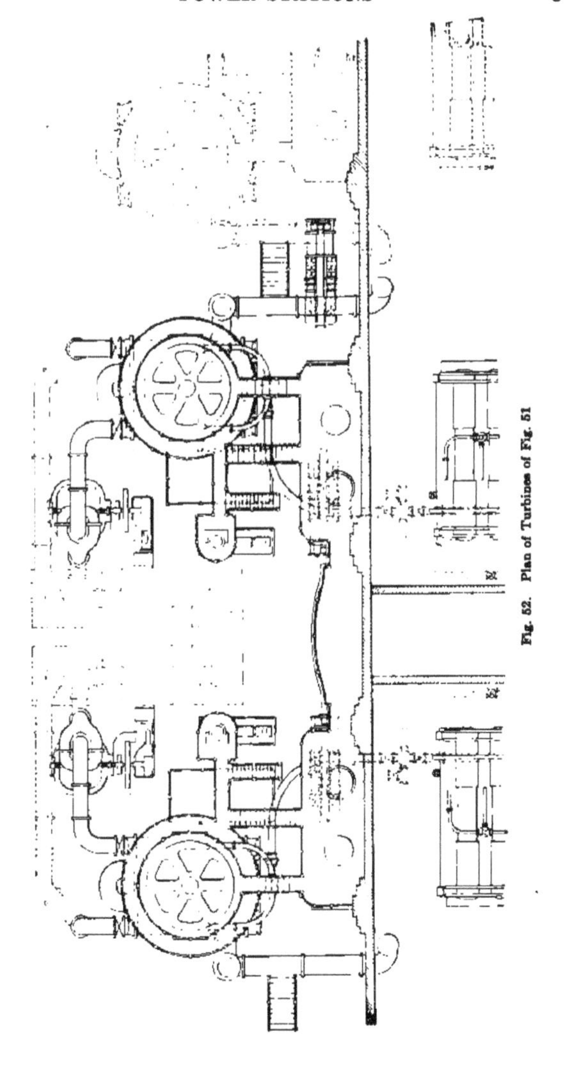

Fig. 52. Plan of Turbines of Fig. 51

should consist of switchboard records, engine-room records, boiler-room records, and distributing-system records. Such records accurately kept and properly plotted in the form of curves, serve admirably for the comparison of station operations from day to day and for the same periods for different years. It pays to keep these records even when additional clerical force must be employed.

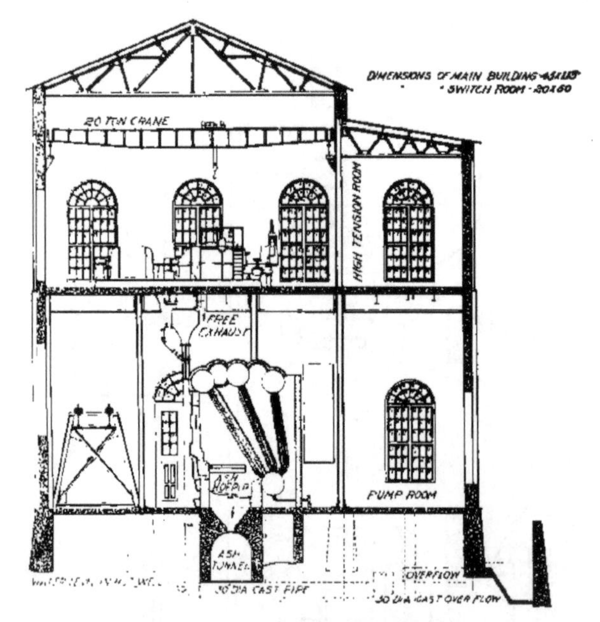

Fig. 53. Double-Deck Turbine Plant

In some states stations furnishing light or power to the public are required to make annual reports and the system of records, accounting, and form of report are all prescribed by the state.

Switchboard records consist, in alternating stations, of daily readings of feeder, recording wattmeters, and total recording wattmeter, together with voltmeter and ammeter readings at intervals of about 15 minutes, in some cases, to check upon the average power factor and determine the general form of the load curve. For direct-current lighting systems, volt and ampere readings serve to give the true output of the stations, and curves are readily plotted from these readings. The voltage should be recorded for the bus bars as well as for the centers of distribution.

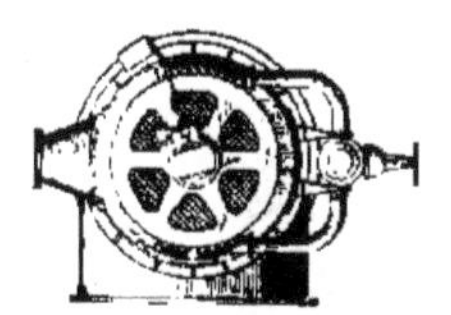

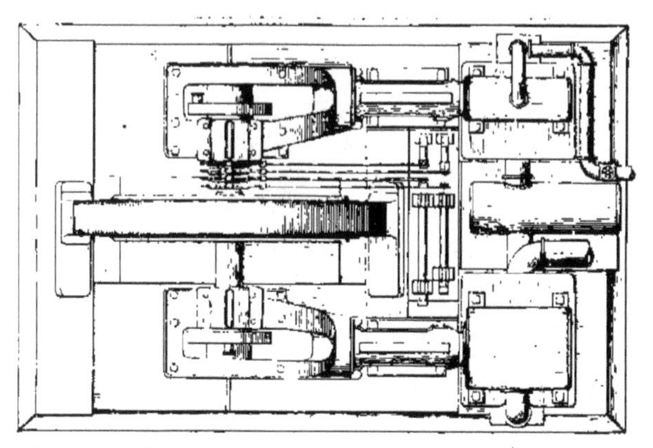

Fig. 54. Space Occupied by Turbo-Alternator Compared with that of Generator and Reciprocating Engine of Same Capacity

Indicator diagrams should be taken from the engines at frequent intervals for the purpose of determining the operation of the valves. Engine-room records include labor; use of waste, oil, and supplies; as well as all repairs made on engines, dynamos and auxiliaries.

Boiler-room records include labor and repairs, amount of coal used, which amount may be kept in detail if desirable, amount of water used, together with steam-gauge record and periodical analysis of flue gases as a check on the methods of firing.

Fig. 55. Arrangement of Units in a Hydraulic Plant

Records for the distributing system include labor and material used for the lines and substations. For multiple-wire systems, frequent readings of the current in the different feeders will serve as a check on the balance of the load.

The cost of generating power varies greatly with the rate at which it is produced as well as upon local conditions. Station-operating expenses include cost of fuel, water, waste, oil, etc., cost of repairs, labor, and superintendence. Fixed charges include insurance, taxes, interest on investment, depreciation, and general office expenses. Total expenses divided by total kilowatt-hours gives

Fig. 56. A Good Arrangement of Apparatus for Substation

the cost of generation of a kilowatt-hour. The cost of distributing a kilowatt hour may be determined in a similar manner. The rate of depreciation of apparatus differs greatly with different machines, but the following figures may be taken as average values, these figures representing percentage of first cost to be charged up each year:

Fireproof buildings from 2 to 3 per cent.
Frame buildings from 5 to 8 per cent.
Dynamos from 2 to 4 per cent.
Prime movers from 2½ to 5 per cent.
Boilers from 4 to 5 per cent.
Overhead lines, best constructed, 5 to 10 per cent.
More poorly constructed lines 20 to 30 per cent.
Badly constructed lines 40 to 60 per cent.
Underground conduits 2 per cent.
Lead covered cables 2 per cent.

Methods of Charging for Power. There are four methods used for charging consumers for electrical energy, namely, *the flat-rate, or contract, system, the meter system, the two-rate meter system,* and *a system by which each customer pays a fixed amount depending on the maximum demand and in addition pays at a reasonable rate for the power actually used.*

In the flat-rate system, each customer pays a certain amount a year for service, this amount being based on the estimated amount of power to be used. These rates vary, depending on the hours of the day during which the power is to be used, being greatest if the energy is to be used during peak hours. It is an unsatisfactory method for lighting service, as many customers are liable to take advantage of the company, burning more lights than contracted for and at different hours, while the honest customer must pay a higher rate than is reasonable in order to make the station operation profitable. This method serves much better when the power is used for driving motors, and is used largely for this class of service.

The simple meter method of charging serves the purpose better for lighting, but the rate here is the same no matter what hour of the day the current is used. Obviously, since machinery is installed to carry the peak of the load, any power used at this time tends to increase the capital outlay from the plant, and users should be required to pay more for the power at such times. The meter system is often employed with a sliding scale or rate, the rate charged per kw.-hour depending upon the amount of electrical energy used.

The two-rate meter accomplishes this purpose to a certain extent. The meters are arranged so that they record at two rates, the higher rate being used during the hours of heavy load.

There are several methods of carrying out the fourth scheme. In the Brighton System a fixed charge is made each month, depending on the maximum demand for power during the previous month, a regular schedule of such charges being made out, based on the cost of the plant. An integrating wattmeter is used to record the energy consumed, while a so-called "demand meter" records the maximum rate of demand.

BIBLIOGRAPHY

Bibliography

THE following list of books on the various subjects of Engineering is not intended to be complete, but represents the best book literature in the field. Unfortunately, on account of lack of space, the many short pamphlets, addresses, and reprints from periodicals, containing a wealth of authoritative information, cannot be included. For these the reader is referred to the annuals and quarterlies of the various engineering organizations and to the libraries.

ABRAHAM-BAYLE: Steam Economy in the Sugar Factory. 96 pp. John Wiley & Sons.

ALLEN-BURSLEY: Heat Engines; steam, gas, steam turbines, and their auxiliaries. A. C. McClurg & Company.

ALLEN, H.: Modern Power Gas-Producer Practice and Application. 344 pp. Van Nostrand Company.

AUCHINCLOSS, W. S.: The Practical Application of the Slide Valve and Link Motion to Stationary, Portable, Locomotive, and Marine Engines. 15th edition, revised. 144 pp. Van Nostrand Company.

BACON, F. W.: A Treatise on the Richards Steam Engine Indicator. 4th edition. 180 pp. Van Nostrand Company.

BALDWIN, WILLIAM J.: Baldwin on Heating. 16th edition, revised and enlarged. 404 pp. John Wiley & Sons.

.... Hot Water Heating and Fitting. The methods of construction and the principles involved. 4th edition, revised and enlarged. 306 pp. McGraw-Hill Company.

.... Heating. Steam heating for buildings, revised. A. C. McClurg & Company.

BALE, M. P.: Pumps and Pumping. 5th edition. 121 pp. Van Nostrand Company.

BALL, R. S.: Natural Sources of Power. 364 pp. Van Nostrand Company.

BARKER, A. H.: Graphic Methods of Engine Design. 2nd edition. 217 pp. Van Nostrand Company.

BARR, WILLIAM M.: Combustion of Coal and the Prevention of Smoke. A. C. McClurg & Company.

BARRUS, G. H.: Boiler Tests. 252 pp. Van Nostrand Company.

.... Engine Tests. 338 pp. Van Nostrand Company.

BARTLETT, F. W.: Mechanical Drawing. 3rd edition, revised. 164 pp. John Wiley & Sons.

BAXTER, WILLIAM, Jr.: Hydraulic Elevators. 300 pp. McGraw-Hill Company.

BEGTRUP, JULIUS: The Slide-Valve and its Functions. With special reference to modern practice in the United States. A. C. McClurg & Company.

BERRY, CHARLES W.: The Temperature-Entropy Diagram. 3rd edition, revised and enlarged. 393 pp. John Wiley & Sons.

BERTIN, L. E.: Marine Boilers. Their Construction and Working. 2nd edition, revised and enlarged. 694 pp. Van Nostrand Company.

BOOTH, W. H.: Superheaters, Superheating, and Their Control. 170 pp. Van Nostrand Company.

.... Water Softening and Treatment. 308 pp. Van Nostrand Company.

BOWKER, WILLIAM R.: Dynamos, Motors, and Switchboard Circuits. Deals with direct, alternating, and polyphase currents. Van Nostrand Company.

BOYCOTT, G. W. M.: Compressed-Air Work and Diving. A handbook for engineers, comprising deep water diving and the use of compressed air for sinking caissons and cylinders and for driving subaqueous tunnels. A. C. McClurg & Company.

BRACKETT, C. F., AND OTHERS: Electricity in Daily Life. A popular account of the application of electricity to everyday use. A. C. McClurg & Company.

BRAGG, E. M.: Marine Engine Design. 175 pp. Van Nostrand Company.

BRANCH, JOSEPH G.: Conversations on Electricity. An elementary book. 282 pp. Rand McNally & Company.

BROMLEY-COBLEIGH: Mathematics for the Practical Engineer. 220 pp. McGraw-Hill Company.

BURNS-BRANCH: Practical Mathematics for the Engineer and Electrician. Covers the mathematics necessary to understand alternating currents. Riley & Sons.

BURNS, ELMER E.: The Electric Motor its Practical Operation. Treats of the principles and operation of all kinds of motors. 191 pp. Numerous tables. Riley & Sons.

CARDULLO, FORREST E.: Practical Thermodynamics. A clear treatment of the natural laws and physical principles which underlie the action of thermodynamic apparatus. 414 pp. McGraw-Hill Company.

CARPENTER-DIEDERICHS: Experimental Engineering and Manual for Testing. 7th edition, rewritten and enlarged. 1132 pp. John Wiley & Sons.

.... Internal-Combustion Engines. Their theory, construction and operation. A. C. McClurg & Company.

CARPENTER, R. C.: Heating and Ventilating Buildings. A manual for heating engineers and architects. A. C. McClurg & Company.

.... The Heating and Ventilation of Buildings. 5th edition, revised. 562 pp. John Wiley & Sons.

CHALKLEY, A. P.: Diesel Engines for Land and Marine Work. 237 pp. Van Nostrand Company.

CHILD, CHARLES T.: The How and Why of Electricity. A book of information for non-technical readers. Van Nostrand Company.

CHRISTIE, W. W.: Chimney Design and Theory. 2nd edition, revised and enlarged. 200 pp. Van Nostrand Company.

.... Water, Its Purification and Use in the Industries. 230 pp. Van Nostrand Company.

CLARK, CARL H.: Marine Gas Engines. Their Construction and Management. 117 pp. Van Nostrand Company.

CLARK, D. K.: Fuel: Its Combustion and Economy. 4th edition. 366 pp. Van Nostrand Company.

CLERK, DUGALD: The Gas and Oil Engine. A. C. McClurg & Company.

.... The Gas, Petrol, and Oil Engine. Thermodynamics of the gas, petrol, and oil engine, together with historical sketch. New and revised edition.

COLLINS, HUBERT E.: Boilers. McGraw-Hill Company.

.... Erecting Work. McGraw-Hill Company.

.... Knocks and Kinks. McGraw-Hill Company.

.... Pipes and Piping. McGraw-Hill Company.

.... Pumps. McGraw-Hill Company.

.... Shaft Governors. McGraw-Hill Company.

.... Shafting, Pulleys, Belting and Rope Transmission. McGraw-Hill Company.

.... Steam Turbines. McGraw-Hill Company.

.... Steam Turbines. A book of instruction for the adjustment and operation of the principal types of this class of prime movers. A. C. McClurg & Company.

.... Valve Setting; simple methods of setting the plain slide valve, Meyer cutoff, Corliss, and Poppet types. A. C. McClurg & Company.

COOLIDGE, C. E.: A Manual of Drawing. 92 pp. John Wiley & Sons.

CREIGHTON, W. H.: Steam Engine and Other Heat Motors. 2nd edition, revised and enlarged. 598 pp. John Wiley & Sons.

CROCKER-ARENDT: Electric Motors. Their action, control, and application. 296 pp. McGraw-Hill Company.

CROCKER-WHEELER: Management of Electrical Machinery. A. C. McClurg & Company.

.... Practical Management of Dynamos and Motors. 206 pp. Spon & Chamberlain.

CROCKER, FRANCIS B.: Electric Lighting. In two volumes. Vol. I: The Generating Plant. Vol. II: Distributing System and Lamps. McGraw-Hill Company.

CROMWELL, J. HOWARD: A Treatise on Belts and Pulleys. 271 pp. John Wiley & Sons.

.... A Treatise on Toothed Gearing. 245 pp. John Wiley & Sons.

CUSHING, H. C., JR.: Standard Wiring for Electric Light and Power. Van Nostrand Company.

DAY, C.: Indicator Diagrams and Engine and Boiler Testing. 4th edition. 220 pp. Van Nostrand Company.

DINGER, H. C.: Handbook for the Care and Operation of Naval Machinery. 2nd edition. 312 pp. Van Nostrand Company.

DRAPER, CHARLES H.: Heat and the Principles of Thermodynamics. New and revised edition. 444 pp. Van Nostrand Company.

DUBBEL, H.: High Power Gas Engines. 200 pp. Van Nostrand Company.

DURAND, W. F.: Practical Marine Engineering. For marine engineers and students, with aids for applicants for marine engineers' licenses. A. C. McClurg & Company.

ELLENWOOD, F. O.: Steam Charts. John Wiley & Sons.

ENNIS, WILLIAM D.: Applied Thermodynamics for Engineers. 3rd edition, revised and enlarged. 514 pp Van Nostrand Company.

EWING, J. A.: The Steam Engine and Other Heat Engines. 3rd edition, revised and enlarged. A. C. McClurg & Company.

FLATHER, J. J.: Rope Driving. 230 pp. John Wiley & Sons.

FOSTER, HORATIO A.: Electrical Engineer's Pocketbook. Tables, data, and formulas relating to all branches of electrical application. 6th edition, completely revised and enlarged. Pocket size. Flexible leather. 1636 pp. 1912. McGraw-Hill Company.

FRANKLIN-ESTY: Elements of Electrical Engineering. Two vols. Van Nostrand Company.

FRENCH, LESTER G.: Steam Turbines. A comprehensive treatment of the whole subject. 418 pp. McGraw-Hill Company.

FRENCH, THOMAS E.: Engineering Drawing. 289 pp. McGraw-Hill Company.

GARCKE-FELLS: Factory Accounts. A new enlarged edition of the standard English treatise on Factory Accounting. 292 pp. McGraw-Hill Company.

GEAR-WILLIAMS: Electric Central Station Distribution Systems. Their design and construction. 352 pp. 1911. Van Nostrand Company.

GEBHARDT, G. F.: Steam Power Plant Engineering. 4th edition, revised and enlarged. 989 pp. John Wiley & Sons.

GILL, AUGUSTUS H.: Engine Room Chemistry. McGraw-Hill Company.

.... Gas and Fuel Analysis for Engineers. 7th edition, revised and enlarged. 141 pp. John Wiley & Sons.

GOING, CHARLES B.: Principles of Industrial Engineering. 174 pp. McGraw-Hill Company.

GOODEVE, T. M.: Textbook on the Steam Engine. 15th edition. 416 pp. Van Nostrand Company.

GOULD, E. S.: The Arithmetic of the Steam Engine. 80 pp. Van Nostrand Company.

GOSS, W. F.: Locomotive Performance. 439 pp. John Wiley & Sons.

.... Locomotive Sparks. 172 pp. John Wiley & Sons.

GREENE, ARTHUR M.: Elements of Heating and Ventilation. 324 pp. John Wiley & Sons.

.... Elements of Refrigeration. John Wiley & Sons.

.... Pumping Machinery. 703 pp. John Wiley & Sons.

GRIMSHAW, R.: Engine Runner's Catechism. Tells how to erect, adjust, and run the principal steam engines in use in the United States; describes the principal features of the different engines. A. C. McClurg & Company.

.... Steam Engine Catechism. Practical answers to practical questions. A. C. McClurg & Company.

GROVER, F.: Practical Treatise on Modern Gas and Oil Engines. 5th edition. 380 pp. Van Nostrand Company.

GULDNER, HUGO: The Design and Construction of Internal-Combustion Engines. 2nd edition, revised and enlarged. A. C. McClurg & Company.

HAEDER, H.: A Handbook on the Steam Engine. 3rd edition, revised. 465 pp. Van Nostrand Company.

HAEDER-HUSKISSON: A Handbook on the Gas Engine. 330 pp. McGraw-Hill Company.

HALL, H. R.: Governors and Governing Mechanism. 2nd edition, enlarged. 188 pp. Van Nostrand Company.

HALSEY, F. A.: Slide Valve Gears. 12th edition, revised and enlarged. 213 pp. Van Nostrand Company.

HARRIS, ELMO G.: Compressed Air. Theory and computations. 123 pp. McGraw-Hill Company.

HARTMAN, FRANCIS M.: Heat and Thermodynamics. 346 pp. McGraw-Hill Company.

HAUSBRAND, E.: Drying by Means of Air and Steam. 77 pp. Van Nostrand Company.

.... Evaporating, Condensing, and Cooling Apparatus. 400 pp. Van Nostrand Company.

HAVEN-SWETT: The Design of Boilers and Pressure Vessels. John Wiley & Sons.

HAWKINS, GEO. W.: Economy Factor in Steam Power Plants. Diagrams, graphs, and tables of the greatest value to the modern power plant. 133 pp. McGraw-Hill Company.

HAWKINS, N.: Handbook of Calculations for Engineers. Relating to the steam engine, the steam boiler, pumps, shafting, etc. A. C. McClurg & Company.

.... Instructions for the Boiler Room. Useful to engineers, firemen, and mechanics. A. C. McClurg & Company.

.... New Catechism of the Steam Engine. A. C. McClurg & Company.

HAWKINS-WALLIS: The Dynamo, Its Theory, Design, and Maintenance. Direct and Alternating Currents. Two vols. A. C. McClurg & Company.

HECK, ROBERT C. H.: The Steam Engine and Other Steam Motors. Two vols. Van Nostrand Company.

.... The Steam Engine and Turbine. 625 pp. Van Nostrand Company.

HEMENWAY, FRANK F.: Indicator Practice and Steam-Engine Economy. 184 pp. John Wiley & Sons.

HENRY-HORA: Modern Electricity. A textbook for students and apprentices. 355 pp. Laird & Lee.

HIRSHFELD-BARNARD: Elements of Heat-Power Engineering. 811 pp. John Wiley & Sons.

HIRSHFELD, C. F.: Engineering Thermodynamics. 2nd edition. 162 pp. Van Nostrand Company.

HIRSHFELD-ULBRICHT: Gas Engines for the Farm. 239 pp. John Wiley & Sons.

.... Gas Power. 209 pp. John Wiley & Sons.

.... Steam Power. John Wiley & Sons.

HISCOX, G. D.: Gas, Gasoline, and Oil Engines. Including producer-gas plants. A. C. McClurg & Company.

HOBART-ELLIS: Armature Construction. 348 pp. Macmillan Company.

HOBART, HENRY M.: Design of Polyphase Generators and Motors. 260 pp. McGraw-Hill Company.

.... Electricity. A textbook designed particularly for engineering students. 226 pp. 43 tables. 1910. Van Nostrand Company.

.... Electric Motors. Continuous, polyphase, and single-phase motors. Their theory and construction. Revised and enlarged. 736 pp. Macmillan Company.

HOFFMAN, JAMES D.: Handbook for Heating and Ventilating Engineers. A practical discussion, with tables and charts, on design and installation. 402 pp. McGraw-Hill Company.

HOGLE, W. M.: Internal Combustion Engines. A reference book for designers, operators, engineers, and students. 256 pp. McGraw-Hill Company.

HOLMES, G. C. V.: Steam Engine. An elementary treatise supplementary to the study of physics. A. C. McClurg & Company.

HOOPER-WELLS: Electric Handbook for Engineering Students. 170 pp. Ginn & Company.

HOPKINS, N. M.: Model Engines and Small Boats. 84 pp. Van Nostrand Company.

HORSTMAN-TOUSLEY: Modern Wiring Diagrams and Designs. 293 pp. F. J. Drake & Company.

.... Practical Armature and Magnet Winding. 230 pp. 1909. A. C. McClurg & Company.

HOUSTON-KENNELLY: The Electric Motor. 377 pp. A. C. McClurg & Company.

HUBBARD, CHARLES L.: Heating and Ventilating Plants. 2nd edition. 300 pp. McGraw-Hill Company.

.... Power, Heating, and Ventilation. A series of three distinct works on the design, construction and management of power, heating, and ventilating plants. McGraw-Hill Company.

.... Steam Power Plants. 2nd edition. 300 pp. McGraw-Hill Company.

HUTTON, F. R.: Heat and Heat-Engines. A study of the principles which underlie the mechanical engineering of a power plant. A. C. McClurg & Company.

.... The Gas Engine. A treatise on the internal-combustion engine using gas, gasoline, kerosene, or other hydrocarbons, as source of energy. A. C. McClurg & Company.

.... The Mechanical Engineering of Power Plants. A. C. McClurg & Company.

.... The Gas Engine. 3rd edition, revised. 562 pp. John Wiley & Sons.

.... Mechanical Engineering of Steam Power plants. 3rd edition, rewritten. John Wiley & Sons.

HUTTON, W. S.: Practical Engineer's Handbook. Comprising a treatise on modern engines and boilers, marine locomotive, and stationary. 7th edition, revised and enlarged. 577 pp. Van Nostrand Company.

.... Steam Boiler Construction. 4th edition, revised and enlarged. 675 pp. Van Nostrand Company.

IBBETSON, W. S.: Practical Electrical Engineering for Elementary Students 167 pp. 1910. Van Nostrand Company.

INNES, C. H.: Air Compressors and Blowing Engines. 300 pp. Van Nostrand Company.

.... Centrifugal Pumps, Turbines and Water Motors. 5th edition, 350 pp. Van Nostrand Company.

.... The Fan: Including the Theory and Practice of Centrifugal and Axial Fans. 258 pp. Van Nostrand Company.

IVENS, EDMUND M.: Pumping By Compressed Air. John Wiley & Sons.

JAMES-DOLE: Mechanism of the Steam Engine and Similar Machines. John Wiley & Sons.

JONES, FORREST R.: The Gas Engine. 447 pp. John Wiley & Sons.

JUNGE, F. E.: Gas Power. The full story of gas power, its generation, transmission, and application, with especial reference to large engines. 548 pp. McGraw-Hill Company.

KELLEY, H. H.: Engineer's Examiner. 160 pp. McGraw-Hill Company.

.... Engine Room Instructor. A collection of the absolutely necessary tables required by the engineer. 80 pp. McGraw-Hill Company.

KENNEDY, R.: Modern Engines and Power Generators. A practical work on prime movers and the transmission of power; steam, electric, water, and hot air. Six vols. Van Nostrand Company.

KENT, WILLIAM: Investigating an Industry. 126 pp. John Wiley & Sons.

.... The Mechanical Engineers' Pocket Book. 8th edition, revised and enlarged. 1461 pp. John Wiley & Sons.

.... Steam-Boiler Economy. A treatise on the theory and practice of fuel economy in the operation of steam boilers. A. C. McClurg & Company.

KEOWN, R. McA.: Mechanism. 170 pp. McGraw-Hill Company.

KERR, E. W.: Power and Power Transmission. Deals with power transmission and generation by the common means other than electrical. A. C. McClurg & Company.

KERSHAW, J. B. C.: Fuel, Water, and Gas Analysis. 167 pp. Van Nostrand Company.

KING, WILLIAM R.: Steam Engineering. 450 pp. John Wiley & Sons.

.... The Elements of the Mechanics of Materials and of Power Transmission. 266 pp. John Wiley & Sons.

KIRSCHKE, A.: Gas and Oil Engines. 160 pp. Van Nostrand Company.

KLEIN, J. F.: Design of a High Speed Steam Engine. 2nd edition, revised and enlarged. 257 pp. Van Nostrand Company.

KNEASS, STRICKLAND L.: Practice and Theory of the Injector. 3rd edition, revised and enlarged. 175 pp. John Wiley & Sons.

KNOX, C. E.: Electric Light Wiring. 219 pp. McGraw-Hill Company.

KOESTER, FRANK: Steam Electric Power Plants. A treatise on the design of central light and power stations, and their economical construction and operation. 2nd edition. 455 pp. 1910. Van Nostrand Company.

LALLIER, ERNEST V.: An Elementary Manual of the Steam Engine. 274 pp. Van Nostrand Company.

LATTA, NISBET: American Producer Gas Practice and Industrial Gas Engineering. 547 pp. Van Nostrand Company.

LEASK, A. R.: Refrigerating Machinery. 4th edition. 296 pp. Van Nostrand Company.

LEDOUX, M.: Ice-Making Machines. The theory of the action of the various forms of cold producing machines. 6th edition, revised. 258 pp. Van Nostrand Company.

LEVIN, A. M.: The Modern Gas Engine and the Gas Producer. 485 pp. John Wiley & Sons.

LEWES, V. B.: Liquid and Gaseous Fuels and the Part They Play in Modern Power Production. 348 pp. Van Nostrand Company.

LINEHAM, WILFRID J.: A Textbook of Mechanical Engineering. 2 parts. A. C. McClurg & Company.

LODGE, WILLIAM: Rules of Management. 139 pp. McGraw-Hill Company.

LOEWENSTEIN, L. C., and CRISSEY, C. P.: Centrifugal Pumps, Their Design and Construction. 432 pp. Van Nostrand Company.

LOW, F. R.: Condensers. 79 pp. McGraw-Hill Company.

.... The Steam Engine Indicator. Direction for the selection, care, and use of the instrument and the analysis and computation of the diagram. A. C. McClurg & Company.

.... The Compound Engine. 52 pp. McGraw-Hill Company.

LUCKE, CHARLES EDWARD: Engineering Thermodynamics. 1176 pp. McGraw-Hill Company.

LUCKE, C. E.: Gas Engine Design. A. C. McClurg & Company.

MACINTIRE, H. J.: Mechanical Refrigeration. A treatise for technical students and engineers. 346 pp. John Wiley & Sons.

MAC CORD, C. W.: Slide Valves. 168 pp. John Wiley & Sons.

.... Velocity Diagrams. Their construction and their uses. John Wiley & Sons.

MAHAN-THOMPSON: Industrial Drawing. Revised and enlarged. 2 vols. 209 pp. John Wiley & Sons.

MARKS-DAVIS: Tables and Diagrams of the Thermal Properties of Saturated and Superheated Steam. A. C. McClurg & Company.

MARKS, E. C. R.: Notes on the Construction and Working of Pumps. 2nd edition, enlarged. 267 pp. Van Nostrand Company.

MARLOW, THOMAS G.: Drying Machinery and Practice. A handbook on the theory and practice of drying and desiccating, with classified description of installation, machinery, and apparatus. 388 pp. Van Nostrand Company.

MARSHALL, W. J., and SANKEY, H. R.: Gas Engines. 293 pp. Van Nostrand Company.

MASON, CHARLES J.: Arithmetic of the Steam Boiler. A reference book showing the practical applications of arithmetic to steam boilers. McGraw-Hill Company.

MATHOT, R. E.: Construction and General Working of Internal Combustion Engines. 576 pp. Van Nostrand Company.

.... Gas Engines and Producer-Gas Plants. A. C. McClurg & Company.

MATTHEWS, F. E.: Elementary Mechanical Refrigeration. 172 pp. McGraw-Hill Company.

MAUJER-BROMLEY: Fuel Economy and Co_2 Recorders. 189 pp. McGraw-Hill Company.

MAYCOCK, W. P.: Electric Wiring, Fittings, Switches, and Lamps. 628 pp. Macmillan Company.

.... Electric Wiring Diagrams. 146 pp. Macmillan Company.

McCULLOCH, R. S.: Elementary Treatise on the Mechanical Theory of Heat, and Its Application to Air and Steam Engines. 288 pp. Van Nostrand Company.

McLOUGHLIN, T. S.: Questions and Answers on National Electrical Code. A key and index to the official code. 232 pp. McGraw-Hill Company.

MEADE, NORMAN G.: Electric Motors, Their Installation, Construction, Operation, and Maintenance. 158 pp. McGraw-Hill Company.

MEIER, KONRAD: Mechanics of Heating and Ventilation. 161 pp. McGraw-Hill Company.

MERRIMAN, MANSFIELD: Mechanics of Materials. 11th edition. 524 pp. John Wiley & Sons.

.... The Strength of Materials. 6th edition, revised and enlarged. 169 pp. John Wiley & Sons.

MEYER, H. C.: Steam Power Plants. Their Design and Construction. A. C. McClurg & Company.

MILLER-BERRY-RILEY: Problems in Thermodynamics and Heat Engineering. 71 pp. John Wiley & Sons.

MONROE, WILLIAM S.: Steam Heating and Ventilation. 140 pp. McGraw-Hill Company.

MOORE, STANLEY H.: Mechanical Engineering and Machine Shop Practice. 502 pp. McGraw-Hill Company.

MORRIS, HENRY H.: An Introduction to the Study of Electrical Engineering. 2nd edition, revised. 418 pp. McGraw-Hill Company.

MORRIS, WILLIAM L.: Steam Power Plant Piping Systems. Design, installation, and operation. 490 pp. McGraw-Hill Company.

MOYER, JAMES A.: Power Plant Testing. A complete treatise on the generally approved methods used for testing engines, turbines, boilers, and the auxiliary machinery found in the average power plant. 2nd edition. 432 pp. McGraw-Hill Company.

.... The Steam Turbine. 2nd edition, revised and enlarged. 430 pp. John Wiley & Sons.

MUNRO-JAMIESON: Pocketbook of Electrical Rules and Tables. 19th edition. Leather. Pocket size. 810 pp. J. B. Lippincott & Company.

NEILSON, R. M.: The Steam Turbine. A. C. McClurg & Company.

PARSHALL-HOBART: Armature Winding for Electric Machines. 3rd edition. 278 pp. Numerous tables. Van Nostrand Company.

PARSONS, H.: Steam Boilers, Their Theory and Design. A. C. McClurg & Company.

PARTINGTON, JAMES R.: A Textbook of Thermodynamics. 550 pp. Van Nostrand Company.

PATERSON, G. W.: The Management of Dynamos. A. C. McClurg & Company.

PAULDING, C. P.: Practical Laws and Data on the Condensation of Steam in Covered and Bare Pipes. 107 pp. Van Nostrand Company.

.... Transmission of Heat Through Cold-Storage Insulation. 41 pp. Van Nostrand Company.

PEABODY, C. H.: Computation for Marine Engines. 209 pp. John Wiley & Sons.

.... A Manual of the Steam Engine Indicator. 153 pp. John Wiley & Sons.

.... Thermodynamics of the Steam Turbine. 282 pp. John Wiley & Sons.

.... Valve Gears for Steam Engines. 2nd edition, revised. 142 pp. John Wiley & Sons.

.... Thermodynamics of Steam Engine and Other Heat Engines. A. C. McClurg & Company.

.... Thermodynamics of the Steam Engine and other Heat Engines. 6th edition, revised. 543 pp. John Wiley & Sons.

PEABODY-MILLER: Steam Boilers. A. C. McClurg & Company.

.... Steam Boilers. 3rd edition, revised and enlarged. 543 pp. John Wiley & Sons.

PEELE, ROBERT: Compressed Air Plant. 2nd edition, revised and enlarged. 502 pp. John Wiley & Sons.

PENDRED, VAUGHAN: The Railway Locomotive. 322 pp. Van Nostrand Company.

PERKINS, HENRY A.: An Introduction to General Thermodynamics. 247 pp. John Wiley & Sons.

PERRINE, F. A. C.: Conductors for Electrical Distribution, Their Materials and Manufacture. Calculation of circuits, pole line construction, underground working, and other uses. 2nd edition, revised. 297 pp. 1907. A. C. McClurg & Company.

PERRY, JOHN: The Steam Engine and Gas and Oil Engines. A book for the use of students who have time to make experiments and calculations. A. C. McClurg & Company.

PICKWORTH, C. N.: The Indicator Handbook. Two parts. Van Nostrand Company.

POCHET, M. LEON: Steam Injectors. Their Theory and Use. A. C. McClurg & Company.

POOLE, CECIL P.: The Gas Engine. 97 pp. McGraw-Hill Company.

.... The Wiring Handbook. 32 complete labor-saving tables and a digest of underwriters' rules. Flexible leather. 85 pp. McGraw-Hill Company.

POOLE, HERMAN: The Calorific Power of Fuels. A. C. McClurg & Company.

.... The Calorific Power of Fuels. 2nd edition, revised and enlarged. 279 pp. John Wiley & Sons.

POPPLEWELL, W. C.: Elementary Treatise on Heat and Heat Engines. 384 pp. Van Nostrand Company.

.... Prevention of Smoke, Combined with the Economical Combustion of Fuel. 220 pp. Van Nostrand Company.

PORTER, CHARLES T.: Engineering Reminiscences. 335 pp. John Wiley & Sons.

POTTER, ANDREY A.: Farm Motors. It gives practical information on construction, working, and management of engines and motors suitable for farm use. 261 pp. McGraw-Hill Company.

PRATT, H. K.: Boiler Draught. 205 pp. Van Nostrand Company.

PRAY, T., JR.: Calorimeter Tables for Steam. 93 pp. Van Nostrand Company.

.... Steam Tables and Engine Constants. 126 pp. Van Nostrand Company.

PULLEN, W. W. F.: Injectors. Theory, Construction, and Working. 3rd edition. 214 pp. Van Nostrand Company.

RANDALL, J. A.: Exercises in Heat. John Wiley & Sons.

.... Heat. 331 pp. John Wiley & Sons.

RANKINE, W. J. M.: A Manual of the Steam Engine and Other Prime Movers. 17th edition, revised. 672 pp. Van Nostrand Company.

REAGAN, H. C.: Locomotives: Simple, Compound, and Electric. 5th edition, revised and enlarged. 932 pp. John Wiley & Sons.

REED'S Marine Boilers. A treatise on the causes and prevention of priming. 3rd edition, rewritten and enlarged. 364 pp. Van Nostrand Company.

REED, STANLEY, J.: Turbines Applied to Marine Propulsion. 182 pp. Van Nostrand Company.

RICHARDS, FRANK: Compressed Air. 203 pp. John Wiley & Sons.

.... Compressed Air Practice. An authoritative treatment of compressed air practice with emphasis on the latest developments, applications, and future possibilities. 330 pp. McGraw-Hill Company.

RICHARDSON, JOHN: The Modern Steam Engine: Theory, Design, Construction, Use; A Practical Treatise. A. C. McClurg & Company.

RIMMER, E. J.: Boiler Explosions, Collapses, and Mishaps. 151 pp. Van Nostrand Company.

RIPPER, WILLIAM: Heat Engines. A. C. McClurg & Company.

.... Steam-Engine Theory and Practice. A. C. McClurg & Company.

ROBERTS, E. W.: The Gas-Engine Handbook. A manual of useful information for the designer and the engineer. A. C. McClurg & Company.

ROBERTSON, L. S.: Water-Tube Boilers. 230 pp. Van Nostrand Company.

ROBINSON, STILLMAN W.: Principles of Mechanism. 309 pp. John Wiley & Sons.

ROBINSON, W.: Gas and Petroleum Engines. A manual for students and engineers. A. C. McClurg & Company.

ROE, JOSEPH WICKHAM: Steam Turbines. Theory, design, and field of operation. 143 pp. McGraw-Hill Company.

ROSE, J.: Key to Engines and Engine Running. 417 pp. Van Nostrand Company.

ROWAN, F. J.: The Practical Physics of the Modern Steam Boiler. A. C. McClurg & Company.

SCHWAMB-MERRILL: Elements of Mechanism. 2nd edition, revised and enlarged. 280 pp. John Wiley & Sons.

SEATON, A. E.: A Manual of Marine Engineering. 17th edition, revised and enlarged. 994 pp. Van Nostrand Company.

SENNETT, R.: The Marine Steam Engine. A. C. McClurg & Company.

SEXTON, A. H.: Fuel and Refractory Materials. 2nd edition, revised. 374 pp. Van Nostrand Company.

SHEALY, E. M.: Heat. Designed to supply the fundamental knowledge necessary for the successful study and understanding of the steam engine, gas engine, refrigerating machine, and air compressor. 265 pp. McGraw-Hill Company.

.... Steam Boilers. 374 pp. McGraw-Hill Company.

SHELDON-HAUSMANN: Alternating-Current Machines. Being the second volume of Dynamo-Electric Machinery. Its construction, design, and operation. 9th edition, completely rewritten. 364 pp. 1911. Van Nostrand Company.

.... Dynamo-Electric Machinery. Its construction, design, and operation. Van Nostrand Company.

SHEPARDSON, GEORGE D.: Electrical Catechism. An introductory treatise on electricity and its uses. A. C. McClurg & Company.

SIMONS, THEODORE: Compressed Air. 173 pp. McGraw-Hill Company.

SINCLAIR, ANGUS: Locomotive Engine Running and Management. 22nd edition, enlarged. 442 pp. John Wiley & Sons.

SLOANE, T. O'CONNOR: Electricity Simplified. Theory and practice of electricity. A. C. McClurg & Company.

.... How to Become a Successful Electrician. The studies, methods of work, field of operation, and ethics of the profession. A. C. McClurg & Company.

SMART, RICHARD A.: A Handbook of Engineering Laboratory Practice. 2nd edition, revised and enlarged. 290 pp. John Wiley & Sons.

SMITH, C. A. M., and WARREN, A. G.: The New Steam Tables. 114 pp. Van Nostrand Company.

SNOW, WALTER E.: Steam-Boiler Practice. In its relation to fuels and their combustion and the economic results obtained with various methods and devices. A. C. McClurg & Company.

SOREL-WOODWARD-PRESTON: Carbureting and Combustion in Alcohol Engines. 269 pp. John Wiley & Sons.

SOTHERN, J. W.: Marine Steam Turbine. A practical description of the Parsons marine turbine as at present constructed, fitted, and run. A. C. McClurg & Company.

SPANGLER, H. W.: Notes on Thermodynamics. 5th edition. 80 pp. John Wiley & Sons.

.... Valve Gears. 2nd edition. 179 pp. John Wiley & Sons.

SPANGLER-GREENE-MARSHALL: Elements of Steam Engineering. 3rd edition, revised. 296 pp. John Wiley & Sons.

STAFF OF TWENTY SPECIALISTS: The Standard Handbook for Electrical Engineers. 3rd edition. Flexible red leather. Thumb indexed. 1,500 pp. McGraw-Hill Company.

STEWART, G.: Modern Steam Traps. 112 pp. Van Nostrand Company.

STODOLA, Dr. A.: Steam Turbines: With an Appendix on Gas Turbines and Future of Heat Engines. A. C. McClurg & Company.

STILLMAN, PAUL: The Steam Engine Indicator and the Improved Manometer, Steam and Vacuum Gages, Their Utility and Application. 90 pp. Van Nostrand Company.

STONE, C. H.: Practical Testing of Gas and Gas Meters. 337 pp. John Wiley & Sons.

STROHM, RUFUS T.: Oil Fuel for Steam Boiler. A clear, simple statement of principles that underlie the burning of oil in the furnaces of stationary steam boilers. 145 pp. Pocket size. McGraw-Hill Company.

SUPPLEE, H. H.: The Gas Turbine; progress in the design and construction of turbines operated by gases of combustion. A. C. McClurg & Company.

SWINGLE, CALVIN, E.: Twentieth Century Handbook for Steam Engineers and Electricians. 148 pp. 1910. F. J. Drake Company.

SWOOPE, C. WALTON: Lessons in Practical Electricity. Principles, experiments, and arithmetical problems. 13th edition, revised and enlarged, with chapter on alternating currents. 507 pp. 1912. Van Nostrand Company.

THOMAS, CARL C.: Steam Turbines. 4th edition, revised and enlarged. 330 pp. John Wiley & Sons.

THORKELSON, H. J.: Air Compression and Transmission. 207 pp. McGraw-Hill Company.

THURSTON, R. H.: A Manual of the Steam Engine. 1,000 pp. John Wiley & Sons.

.... A Manual of Steam Boilers, Their Design, Construction, and Operation. 7th edition, revised. 883 pp. John Wiley & Sons.

.... A Manual of the Steam Engine. 2 parts. A. C. McClurg & Company.

.... Treatise on Friction and Lost Work in Machinery and Mill Work. 7th edition. 430 pp. John Wiley & Sons.

TREVERT, EDWARD: Dynamos and Electric Motors. A. C. McClurg & Company.

TULLEY, HENRY C.: Handbook on Engineering, Steam and Electrical. Revised edition. Bound in leather. 1,000 pp. McGraw-Hill Company.

WALKER, SYDNEY F.: A Pocketbook of Electric Lighting and Heating. Comprising formulas, tables, data, and particulars of apparatus for use of central station engineers and contractors. 438 pp. Flexible leather. Henley Company.

.... Steam Boilers, Engines, and Turbines. 410 pp. Van Nostrand Company.

WALLIS-TAYLOR, A. J.: Bearings and Lubrication. 216 pp. Van Nostrand Company.

.... Refrigeration, Cold Storage, and Ice-Making. A practical treatise on the art and science of refrigeration. 3rd edition, revised. 610 pp. Van Nostrand Company.

WATERBURY, L. A.: Laboratory Manual for the Use of Students in Testing Materials of Construction. 270 pp. John Wiley & Sons.

.... A Vest-Pocket Handbook of Mathematics for Engineers. 91 pp. John Wiley & Sons.

WATSON, A. E.: How to Build a Fifty-Light Dynamo or Four-Horsepower Motor. 42 pp. Van Nostrand Company.

.... How to Build a One-Fourth Horsepower Motor or Dynamo. Van Nostrand Company.

.... How to Build a One-Half Horsepower Motor or Dynamo. 32 pp. Van Nostrand Company.

WATSON, E. P.: How to Run Engines and Boilers. With a new section on water-tube boilers. A. C. McClurg & Company.

.... Small Engines and Boilers. 126 pp. Van Nostrand Company.

WEHRENFENNIG-PATTERSON: Analysis and Softening of Boiler Feed Water. 2nd edition, revised. 290 pp. John Wiley & Sons.

WEINGREEN, J.: Electrical Power Plant Engineering. 431 pp. McGraw-Hill Company.

WEISBACH-DU BOIS: Heat, Steam, and Steam Engines. 559 pp. John Wiley & Sons.

WEISBACH-HERRMANN-KLEIN: Kinematics and the Power of Transmission. 544 pp. John Wiley & Sons.

.... Machinery of Transmission and Governors. 549 pp. John Wiley & Sons.

WEISBACH, J., and HERRMANN, G.: Mechanics of Air Machinery. 213 pp. Van Nostrand Company.

WHITHAM, JAY M.: Steam-Engine Design. For the use of mechanical engineers, students, and draftsmen. A. C. McClurg & Company.

WHITMAN, W. G.: Heat and Light in the Household. John Wiley & Sons.

WILDA, H.: Steam Turbines. 203 pp. Van Nostrand Company.

WILLIAMS-TWEEDY: Commercial Engineering for Central Stations. 142 pp. McGraw-Hill Company.

WILSON, THOMAS: Steam Traps. 99 pp. McGraw-Hill Company.

WIMPERIS, H. E.: The Internal Combustion Engine. Being a textbook on gas, oil, and petrol engines. A. C. McClurg & Company.

.... The Principles of the Application of Power to Road Transport. 146 pp. Van Nostrand Company.

WITTBECKER, WILLIAM A.: Domestic Electrical Work. On how to wire buildings for bells, alarms, annunciators, and for gas lighting from batteries. A. C. McClurg & Company.

WOOD, DE VOLSON: Thermodynamics, Heat Motors, and Refrigerating Machines. 8th edition, revised and enlarged. 475 pp. John Wiley & Sons.

.... Turbines. 2nd edition, revised and enlarged. 149 pp. John Wiley & Sons.

WOOD, M. P.: Rustless Coatings; Corrosion and Electrolysis of Iron and Steel. 432 pp. John Wiley & Sons.

WOOLEY-MEREDITH: Shop Sketching. 104 pp. McGraw-Hill Company.

WORDINGHAM, CHARLES H.: Central Electric Stations, Their Design, Organization, and Management. 2nd edition, revised. 496 pp. J. B. Lippincott & Company.

WRIGHT, F. W.: The Design of Condensing Plant. 210 pp. Van Nostrand Company.

ZEUNER, A.: Technical Thermodynamics. 2nd edition. Two vols. Van Nostrand Company.

REVIEW QUESTIONS

REVIEW QUESTIONS

ON THE SUBJECT OF

MANAGEMENT OF DYNAMO-ELECTRIC MACHINERY

PART I

1. What items are to be considered in selecting a machine?
2. What means are applied for connecting the engine, or other prime movers, with the generator?
3. State the formula for determining the approximate width of a single belt required to transmit a given horsepower.
4. What are the advantages of rope driving?
5. What is the usual speed of shafting employed in textile mills?
6. Give diagram for connecting shunt type of direct-current generator.
7. Explain how to run two compound-wound generators in parallel.
8. What is the chief difference in appearance of a synchronous converter as compared with a direct-current generator?
9. Considering a six-phase rotary converter, the alternating current across phases 1 and 2 is 1000 volts. What is the direct current?
10. Give wiring connection for two direct-current generators on a three-wire system.
11. How is the speed control of a shunt motor obtained?
12. What is a differentially-wound motor?
13. Considering a three-phase generator Y connected, what is the value of the line current in comparison with the current per phase?
14. How is the direction of rotation of induction motors reversed?
15. Give the characteristics of repulsion induction motors.

REVIEW QUESTIONS

ON THE SUBJECT OF

MANAGEMENT OF DYNAMO-ELECTRIC MACHINERY

PART II

---

1. How can the friction of the bearings and brushes be roughly tested?

2. Which is the most accurate way to determine the temperature rise in electrical apparatus?

3. How is the balance tested on small and medium-sized machines?

4. What is the safe-carrying capacity of #5 B & S rubber insulated copper wire?

5. Make a sketch of an enclosed fuse.

6. Describe the fall-of-potential method for locating faults.

7. Describe the voltmeter test for insulation resistance.

8. For what reasons are water-box resistances used?

9. What is a tachometer?

10. Calculate the torque of a machine, the horsepower of which equals 50 and its speed is 550 revolutions per minute at full load.

11. According to what formula can the mechanical power of a generator or motor be calculated?

12. Give an expression for the efficiency of a generator and for the efficiency of a motor.

13. In which cases are instrument transformers employed.

14. Which is the most common trouble encountered with dynamo-electric machinery?

15. To what is the humming noise in an alternator due?

16. What are the general causes for a motor to stop or to fail to start?

REVIEW QUESTIONS

ON THE SUBJECT OF

TYPES OF GENERATORS AND MOTORS

PART I

1. Define dynamo-electric machinery.
2. What are the most important subdivisions of dynamo-electric machinery?
3. In what combinations are direct-current generators manufactured?
4. In what different ways are steam or gas-driven engine generators arranged?
5. What shaft arrangements have water-wheel-driven generators in sizes from about 100 to 10000 kw. capacity?
6. Describe a pulley on a belt-driven generator on a 150-kw. generator.
7. Is a steam turbine ever belted to a generator?
8. In what cases is gearing used for driving generators?
9. What is the best present practice for driving generators?
10. What are the advantages and the disadvantages of direct-driven generators?
11. What should be the distance between the centers of the engine and generator shaft in belt-driven generators?
12. Is there any difference in driving generators or motors?
13. What subdivisions may be made for direct-current machinery?
14. How are generators of sizes above 200 kw. commonly driven?
15. What is the speed of direct-current generators of larger sizes?
16. What is, in general, the relation between the kw. capacity and the speed?

REVIEW QUESTIONS

ON THE SUBJECT OF

TYPES OF GENERATORS AND MOTORS

PART II

1. Give the general classification of alternating-current dynamo electric machinery.

2. What are the standard frequencies of alternators?

3. What are the speed limits of belted alternators; of turbo-alternators?

4. Into what types are alternators divided?

5. Describe the method of ventilation in a Crocker-Wheeler turboalternator.

6. Give a description of the rotor construction of a Westinghouse water-wheel-driven alternator.

7. How are alternating-current motors classified?

8. Give the characteristics of a polyphase induction motor.

9. What is a compensated repulsion motor?

10. State the differences between the AN and ANY type of Allis-Chalmers motors.

11. Describe the single-phase induction motors of the Century Electric Company.

12. Describe and sketch the magnetic bridge as used in the polyphase induction motors of the Crocker-Wheeler Company.

13. Upon what principle does the starting of the single-phase induction motors of the Emerson Electric Manufacturing Company depend?

14. State the differences in Forms K, L, and M of the General Electric Company's polyphase induction motors.

15. Give a description of the performance of a compensated repulsion induction motor of the General Electric Company.

REVIEW QUESTIONS

ON THE SUBJECT OF

POWER STATIONS.

1. Why is it desirable to allow a reserve capacity when deciding upon the number of units to be installed?

2. Draw a diagram of the piping system for what you consider a first-class arrangement of the units.

3. Would you use a high or low-speed engine (*a*) in the case of a 200 K.W. generator; (*b*) in the case of a 300 K.W. generator; (*c*) in the case of a 1,000 K.W. generator?

4. When alternators are driven in parallel by gas engines, what arrangement is used to compensate for the variation in angular velocity?

5. Is it better to have a machine of very high efficiency at the expense of excessive cost, or to have it reasonable in price with lower efficiency?

6. What arrangement is now frequently used to avoid placing the different pieces of apparatus directly on the panels of a switchboard?

7. Explain the function of a substation and tell of what its equipment should consist.

8. What are the advantages of oil switches?

9. Explain why the capacity of a 6-phase rotary converter is greater than that of a single-phase, two-phase, or three-phase.

10. Which system of charging for power do you consider the fairest and best? Explain why.

11. Explain what is meant by a reverse-current relay, and give an example of its usefulness.

12. What factors must be taken into consideration when locating a rotary converter substation?

GENERAL INDEX

GENERAL INDEX

In this Index the Volume number appears in Roman numerals—thus, I, II, III, IV, etc., and the Page number in Arabic numerals—thus, 1, 2, 3, 4, etc. For example, Volume IV, Page 327, is written, IV, 327.

The page numbers of this volume will be found at the bottom of the pages; the numbers at the top refer only to the section.

Note.—For page numbers see foot of pages.

B

Note.—For page numbers see foot of pages.

Note.—For page numbers see foot of pages.

Note.—For page numbers see foot of pages.

C

Note.—For page numbers see foot of pages.

Note.—For page numbers see foot of pages.

Note.—For page numbers see foot of pages.

Note.—For page numbers see foot of pages.

Note.—For page numbers see foot of pages.

Note.—For page numbers see foot of pages.

Note.—For page numbers see foot of pages.

Note.—For page numbers see foot of pages.

Note.—For page numbers see foot of pages.

Note.—For page numbers see foot of pages.

Note.—For page numbers see foot of pages.

Note.—For page numbers see foot of pages.

Note.—For page numbers see foot of pages.

Note.—For page numbers see foot of pages.

Note.—For page numbers see foot of pages.

Note.—For page numbers see foot of pages.

Note.—For page numbers see foot of pages.

Note.—For page numbers see foot of pages.

Note.—For page numbers see foot of pages.

Note.—For page numbers see foot of pages.

Note.—For page numbers see foot of pages.

Note.—For page numbers see foot of pages.

Note.—For page numbers see foot of pages.

Note.—For page numbers see foot of pages.

Note.—For page numbers see foot of pages.